Calvin Cutter

First Book on Analytic Anatomy, Physiology and Hygiene

human and comparative - for common and grammar schools and for families

Calvin Cutter

First Book on Analytic Anatomy, Physiology and Hygiene
human and comparative - for common and grammar schools and for families

ISBN/EAN: 9783337369842

Printed in Europe, USA, Canada, Australia, Japan

Cover: Foto ©berggeist007 / pixelio.de

More available books at **www.hansebooks.com**

ON

ANALYTIC

Anatomy, Physiology and Hygiene,

HUMAN AND COMPARATIVE.

FOR COMMON AND GRAMMAR SCHOOLS AND FOR FAMILIES.

By CALVIN CUTTER, A.M., M.D.,

AUTHOR OF "SECOND BOOK ON ANALYTIC ANATOMY, PHYSIOLOGY AND HYGIENE," "NEW ANALYTIC ANATOMY, PHYSIOLOGY AND HYGIENE," AND "NEW OUTLINE ZOOLOGICAL CHARTS, HUMAN AND COMPARATIVE."

WITH NUMEROUS ENGRAVINGS.

PHILADELPHIA
J. B. LIPPINCOTT & CO.

Entered according to Act of Congress, in the year 1872, by
CALVIN CUTTER, A.M., M.D.,
in the Office of the Librarian of Congress, at Washington.

Copyright also secured in Great Britain and entered at Stationers' Hall, London, and right of Translation reserved.

LIPPINCOTT'S PRESS,
PHILADELPHIA

PREFACE.

This manual, designed for Common and Grammar Schools and for families, has been prepared with several objects in view. First: To so limit its size, to make its analysis so complete, to secure so natural an arrangement, with so apposite and artistic illustrations, as to render it usable in Common Schools having terms of ten or fourteen weeks, and also to present such topics for consideration as shall interest and profit pupils, parents and the people in the laws of health, which are based upon a knowledge of the laws of physiology.

Second: To secure these results, the outlines of the human system are discussed relative to *structure*, to *use* and to *health*, with so much of Comparative Anatomy (Zoology) as is deemed needful to show the analogy that exists between man and the inferior animals, and to appetize for more extended study in the science of life.

The treatment of some of the common ills of life, of wounded, poisoned and asphyxiated persons, is briefly discussed in their appropriate chapters.

Another object in view has been to make the style clear, concise and adapted to the references made to illustrative figures, to teach the pupil the correct word, though it may be a technical term, believing it is best to use those terms which express the ideas that are peculiar to the study. As the study of *objects* is more simple and impressive than mere words, and as illustrations are more instructive, particularly to children, than written sentences, this work has been so arranged as to be used advantageously, with OBJECT study and TOPICAL instruction, especially with Outline Anatomical Charts, both human and comparative.

As *use* is the test of a text-book, this manual is respectfully submitted to teachers and to patrons of schools.

CALVIN CUTTER.

WARREN, MASS., July, 1872

ANALYSIS OF CONTENTS.

CHAPTER I.

SECT.		PAGE
1.	GENERAL REMARKS	7

DIVISION I.

MOTORY APPARATUS.

CHAPTER II.
THE BONES.

2. Anatomy of the Bones.................... 13
3. Physiology " " 20
4. Hygiene " " 22
5. Comparative Osteology.................... 24

CHAPTER III.
THE MUSCLES.

6. Anatomy of the Muscles................. 36
7. Physiology " " 39
8. Hygiene " " 42
9. Comparative Myology...................... 48

DIVISION II.

NUTRITIVE APPARATUS.

CHAPTER IV.
THE DIGESTIVE ORGANS.

10. Anatomy of the Digestive Organs.... 51
11. Physiology " " " 57
12. Hygiene " " " 59
13. Comparative Splanchnology............ 65

CHAPTER V.
ABSORPTION.

14. Anatomy of the Absorbents............ 75
15. Physiology " " 78
16. Hygiene " " 80

CHAPTER VI.
THE RESPIRATORY AND VOCAL ORGANS.

17. Anatomy of the Respiratory and Vocal Organs............................ 82
18. Physiology of the Respiratory and Vocal Organs............................ 86
19. Hygiene of the Respiratory and Vocal Organs............................ 90
20. Comparative Pneumonology............ 96

CHAPTER VII.
THE SKIN.

SECT.		PAGE
21. Anatomy of the Skin....................		102
22. Physiology " "		107
23. Hygiene " "		109

CHAPTER VIII.
THE CIRCULATORY ORGANS.

24. Anatomy of the Circulatory Organs 115
25. Physiology " " " 121
26. Hygiene " " " 122
27. Comparative Angiology................ 124

CHAPTER IX.
ASSIMILATION.

28. Assimilation, General and Special... 129

DIVISION III.

SENSORIAL APPARATUS.

CHAPTER X.
NERVOUS SYSTEM.

29. Anatomy of the Nervous System... 134
30. Physiology " " " ... 140
31. Hygiene " " " ... 142
32. Comparative Neurology................... 148

CHAPTER XI.
SPECIAL SENSES.

33, 34, 35. Sense of Smell.................... 154
36, 37, 38. Sense of Sight.................... 156
39, 40, 41. Sense of Hearing............... 162
42, 43. Sense of Taste........................ 167
44. Sense of Touch............................. 168

APPENDIX.

CHAPTER XII.

CARE OF THE SICK................................ 171
TREATMENT OF WOUNDS...................... 176
" OF BURNS AND SCALDS....... 178
" OF FROST-BITE AND CHIL-
BLAIN.............................. 179
" OF PERSONS APPARENTLY
DROWNED 180
POISONS AND THEIR ANTIDOTES............ 181
GLOSSARY... 185
INDEX 195

ANATOMY, PHYSIOLOGY AND HYGIENE.

CHAPTER I.

GENERAL REMARKS.

§ **1. THE HUMAN BODY AND A MACHINE COMPARED.**—*Division of Objects in Nature. Definitions of Terms. Cells — Tissues — Membranes. Life of Organized Bodies. Great Divisions of the Body.*

1. HOWEVER complete a machine of human invention, none can be more perfect in structure, beautiful in appearance, or harmonious in action than the "house we live in."

A WATCH, for instance, contains beautiful wheels as well as delicate springs, all of which are surrounded by well-fitted cases. Yet the human body contains parts more beautiful, organs more delicate, enclosed in cases more perfect in construction.

The watch has not within itself the power of making or applying the oil necessary for its movable parts, but God in his goodness has so made the parts of the human body that they make and apply as they need their own oily fluid.

The form and size of a watch do not of themselves change; but in form and size man varies from his cradle to his grave. Growth and decay are constant in the human frame.

If a watch is injured, it has not power to repair or mend itself. Not so with the human body: you may bruise it, and the injured part possesses a power that is generally able to heal it.

2. All objects in the material world are divided into *Organic*, as, animals and plants; and *Inorganic*, as, minerals, earths, water and air.

QUESTIONS.—State the comparison between a watch and the human body—The oiling of the parts—The form and size contrasted. What is constant in the body? State the comparison when injuries are received. How are all objects in nature divided?

3. Every organized body is composed of various parts or *Or'gans*. A collection of organs so arranged that their combined actions shall produce a given result is called an *Appara'tus*. The definite, peculiar use of an organ or apparatus is called its *Function*, as, the digestive apparatus consists of the organs—teeth, stomach, liver, etc.—whose combined functions result in the digestion of food.

4. The description of the form and position of these organs is called ANAT'OMY;* the description of their functions, PHYSIOL'OGY;† the examination of the conditions most favorable to their health, HY'GIENE.‡

5. Anatomy, Physiology and Hygiene may be considered as of two kinds, *Human* and *Comparative*. The first pertains to man, the latter to the inferior, or other animals than man; as the horse, the whale.

6. The greatest variety prevails in the organization of different animals. In some the functions are simple. In others they are complex, and generally the more varied the functions are in any animal, the more complex will be its structure.

FIG. 2.

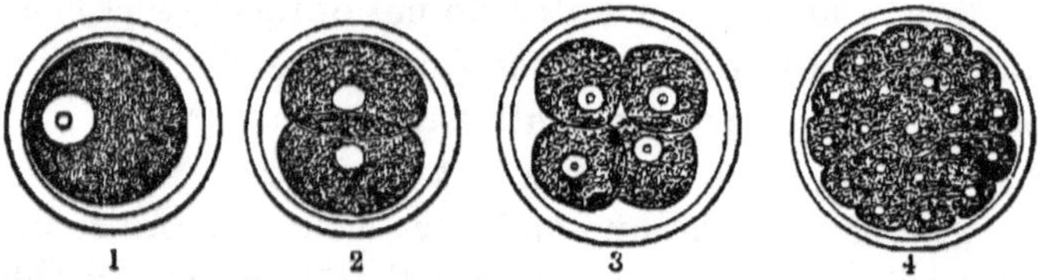

FIG. 2. AN IDEAL CELL.—1, Cell, with its wall, fluid, nucleus and its nucleolus. 2, The same divided into two. 3, The same divided into four cells. 4, The same divided into many cells. The dark portion, the fluid; the white spot, the nucleus; the inner small circle, the nucleolus. Magnified.

Observation.—A good example of a simple animal cell on a large scale is an egg; the lining of the shell is the cell-wall or sac; the white is the contained fluid; the yolk is the *nucleus;* and its germ-spot is the *nucleolus.* (Fig. 2.)

What is an organ? An apparatus? What is function? Define Anatomy—Physiology—Hygiene. How may Anatomy, Physiology and Hygiene be considered? Of what does Human Anatomy treat? Comparative? What is said of the organization of different animals? What is the earliest form of any living thing?

* Gr., *ana*, through, and *tomē*, a cutting.
† Gr., *phusis*, nature, and *logos*, a discourse.
‡ Gr., *hugieinon*, health.

7. The earliest organic form of any living thing is a *Cell*; and cells differently combined form *Tissues*.

8. A SIMPLE CELL consists of a delicate sac containing a fluid, in which is another very minute sac, called the *nu'cleus*,* which contains yet another sac—the *nucleolus* or little nucleus. (Fig. 2.) Very minute particles or granules are also seen.

9. The TISSUES, which, combined in various proportions, make up the organs of the body. The principal tissues of animals are the *Muscular, Nervous* and *Cellular.*

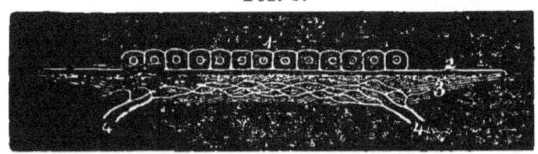

FIG. 3. DIAGRAM EXHIBITING THE RELATIVE POSITION OF THE ELEMENTS OF SEROUS AND MUCOUS MEMBRANES.—1, Epithelium, composed of nucleated cells. 2, Basement layer. 3, Fibrous layer, in which the arteries and veins (4) terminate in a capillary network. Magnified.

10. The simplest of all the tissues, resembling a film of transparent gelatin (jelly), is called *Basement* membrane. (Fig. 3.) Upon it, in various parts of the body, are imbedded minute *epithe'lial* † cells. Other membranes are called the *Se'rous, Syno'vial* and *Mucous.*

The SEROUS MEMBRANE is that portion which lines the walls of certain closed sacs or cavities. It secretes a fluid called *Serum.* The SYNOVIAL MEMBRANE closely resembles the Serous as regards structure. It secretes a fluid called *Synovia.* The MUCOUS MEMBRANE opens to the surface. It secretes a fluid called *Mucus.* This membrane varies in thickness in different parts of the body.

11. All organized bodies have a limited period of life, and

What does a combination of cells form? Of what does a simple cell consist? What builds up the organs of the body? Name some of the tissues. What is Basement Membrane? Serous? Synovial? Mucus?

* L., *nut*. † Gr., *epi*, upon, and *thēlē*, a nipple

this period varies with every species. The duration of some plants is limited to a single summer, while several kinds of

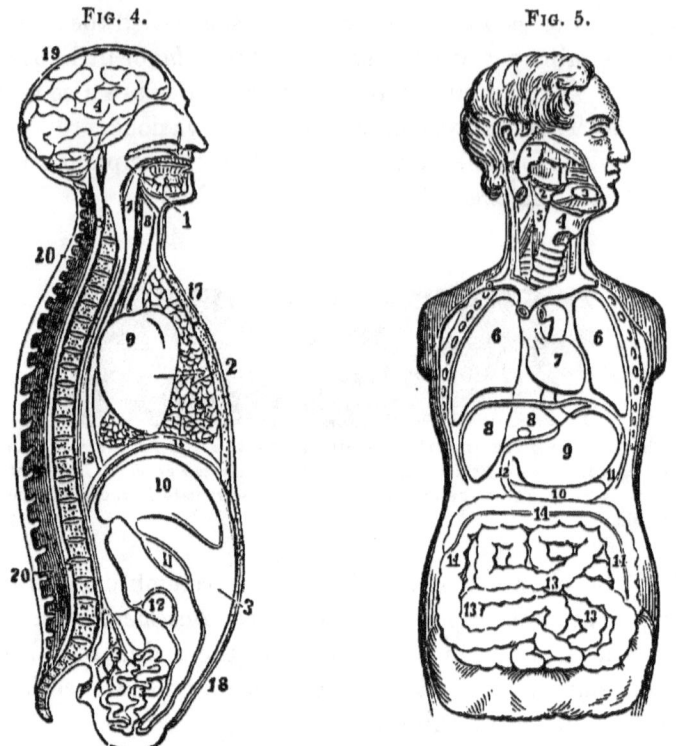

FIG. 4. FIG. 5.

FIG. 4. A SIDE VIEW OF THE TWO GREAT CAVITIES OF THE BODY AND THEIR ORGANS.—1, 2, 3, The *lower* cavity. 1, The mouth. 2, The thorax. 3, The abdomen. (6, A salivary gland. 7, The œsophagus. 8, 8, The trachea and lungs. 9, The heart. 10, The liver. 11, The stomach. 12, The colon. 13, The small intestine. 14, The lacteals. 15, The thoracic duct. 16, The diaphragm.) 17, 18, 20, 20, The walls of the lower cavity, composed of bones, muscles and skin. 4, 5, 5, The *upper* cavity and its organs. 4, The brain. 5, 5, The spinal cord. 19, 20, 20, The walls of the upper cavity. 20, 20, The spinal column.

FIG. 5. THE POSITION OF THE ORGANS OF THE MOUTH, THORAX, AND ABDOMEN.—1, 2, 3, Salivary glands. 4, The larynx and trachea. 5, The œsophagus. 6, 6, The lungs. 7, The heart. 8, 8, The liver. 9, The stomach. 10, The pancreas. 11, The spleen. 13, 13, 13, The small intestine. 14, 14, 14, The large intestine.

trees live many hundred years. Some animals live but a short time, while others live more than a century.

What is said of the limit of life in plants and animals?

12. The life of man is shortened by disease; but disease is under the control of fixed laws—laws which we are capable of understanding and obeying. How important, then, is the study of Physiology and Hygiene!

13. To understand the structure or use of a machine, it is necessary to examine the different parts separately as well as combined. The same is true of the animal frame, so "fearfully and wonderfully made."

14. The human body has two great Cavities: the *Lower* and the *Upper*. (Fig. 4.) The same division is applied to the horse, to birds and to fishes.

The LOWER (Anterior) CAVITY contains the parts of the *Mouth*, *Tho′rax* (Chest), and the *Abdo′men*.

The UPPER (Posterior) CAVITY encloses the *Brain* and the *Spinal Cord*. These great cavities are protected by walls built up of bones, muscles (lean meat), and the whole is covered by the skin.

How is life usually shortened? The importance of the study of Anatomy, Physiology and Hygiene. Name the two great cavities of the body. What does the lower cavity contain? The upper?

SYNTHETIC TOPICAL REVIEW.
GENERAL REMARKS. GENERAL ANALYSIS.
Cells, Tissues, Organs, Apparatus, Divisions

DIVISION I.
MOTORY APPARATUS.

15. In every movement of the body certain organs are brought into action, which, taken collectively, constitute the MOTORY APPARATUS. The parts of this apparatus are the *Bones* and *Joints*, the *Muscles* and the *Nerves of Motion*.

CHAPTER II.
THE BONES.

16. The bones are firm and hard, and of a dull white color. In all the higher orders of animals, among which is man, they

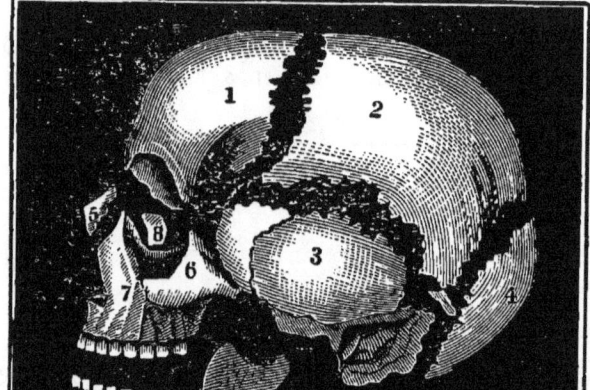

Fig. 6.

Fig. 6. Bones of the Head.—1, Frontal bone. 2, Parietal bone. 3, Temporal bone. 4, Occipital bone. 5, Nasal bone. 6, Malar bone. 7, Upper jaw. 8, Os unguis. 9, Lower jaw.

are in the interior of the body, while in lobsters, crabs, etc., they are on the outside, forming a case, which protects the movable parts from injury.

Name the organs of the Motory Apparatus. Give the structure of bones. Where are they found in man? In lobsters?

§ **2.** ANATOMY OF THE BONES.—*Number and Classification of the Bones. Bones of the Head—Of the Trunk—Of the Lower Extremities—Of the Upper Extremities. The Joints—Definition and Classification. Immovable Joints—Mixed—Movable. Cartilage. Synovial Membrane. Ligaments. Formation of Bone. Periosteum.*

17. The number of bones in the human body exceeds two hundred. These, for convenience, are divided into four parts: 1st The bones of the *Head.* 2d. The bones of the *Trunk.* 3d The bones of the *Lower Extremities.* 4th. The bones of the *Upper Extremities.*

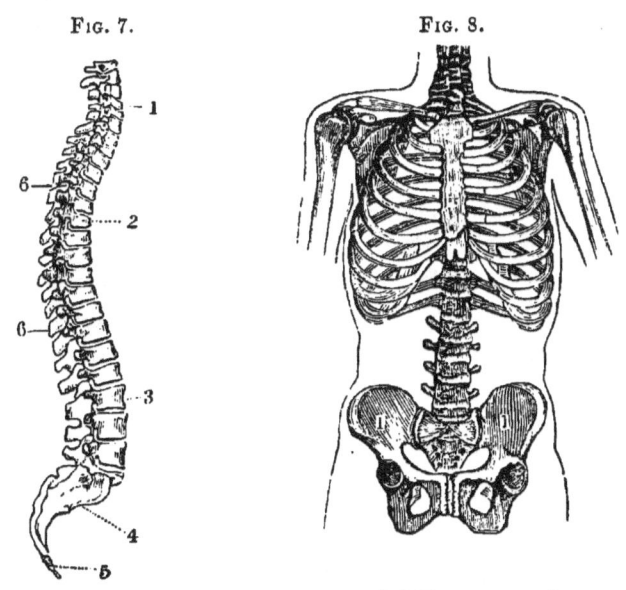

FIG. 7. THE SPINAL COLUMN.—1, 2, 3, Vertebræ. 4, 5, The sacrum and coccyx bones of the pelvis. 6, 6, Processes.

FIG. 8. THE CHEST AND PELVIS.—1, 1, The hip-bones.

18. The BONES OF THE HEAD are divided into those of the *Skull, Ear* and *Face.*

The SKULL is formed of eight bones. These are joined together by ragged edges, called *sut'ures.* (Fig. 6.)

In each EAR are three small bones, which aid in hearing.

How many bones in the human body? How divided? Name the divisions of the bones of the Head. What is said of the Skull-bones? The bones of the Ear?

In the FACE are fourteen bones. They support the softer parts outside of them. (Fig. 6.)

19. The TRUNK has twenty-four *Ribs;* twenty-four bones in the *Spinal Column* (back-bone); the *Sternum* (breast-bone); four in the *Pelvis*, and one at the root of the tongue. (Fig. 8.)

The RIBS are slender bones joined to the spinal column at one extremity, and to the sternum at the other by means of a yielding substance called *Car'tilage* (gristle). The seven upper ribs are united to the sternum by separate cartilages, and are named *true* ribs. The next three are remotely connected to it by long cartilages; these are called *false* ribs. The lowest two are only joined to the spinal column; these are the *floating ribs*.

The ribs, sternum and spinal column form the *Thorax* (chest). This cavity encloses the heart and the lungs. The shape of the chest is conical; the lower part or base should be broader and fuller than the upper part.

The SPINAL COLUMN has twenty-four short pieces of bone (*Vert'ebra*), with sharp *processes* or spines, placed one upon another like a pile or column. Between each vertebra is a thick piece of cartilage that is elastic like India rubber. This yielding substance not only unites the vertebræ, but also allows considerable freedom of motion. (Fig. 7.)

The processes are so arranged that an opening is formed in each vertebra. These bones, coming directly over each other, form a bony canal, in which the *Spinal Cord* is lodged.

The STERNUM is situated in the middle line of the front of the chest, and is held in place chiefly by the ribs. (Fig. 1.)

20. The PELVIS is composed of the INNOMINATUM (hip-bones), the SACRUM and the COCCYX. (Fig. 8.)

The INNOMINATUM is irregularly shaped. Each hip-bone presents the largest surface of any bone in the body. Within

---

The bones of the Face? State the number and names of the bones of the Trunk. Describe the Ribs. Distinguish between true and false ribs. How is the Thorax formed? What is said of the lowest two ribs? Of what is the Spinal Column composed? What of the arrangement of the processes? Where is the Sternum situated? Of what is the Pelvis composed? Describe the Innominatum.

these bones are deep sockets, lined with cartilage, for the reception of the head of the thigh-bone. (Fig. 8.)

The SACRUM is a wedge-shaped bone between the hip-bones. It is the basis of the spinal column. (Fig. 7.)

The COCCYX, at the lower extremity of the spinal column, varies at different ages. In infancy it is cartilaginous; in after life it becomes bony.

21. The LOWER EXTREMITIES contain sixty bones: the *Fe'mur* (thigh-bone); the *Pa-tel'la* (knee-pan); the *Tib'i-a* (shin-bone); the *Fib'u-la* (small bone of the leg); and the bones of the *Foot*.

The FEMUR* is the strongest and longest bone of the body. It supports the weight of the head, trunk and upper extremities. (Fig. 1.)

The PATELLA† is a small chestnut-shaped bone, placed on the front part of the lower extremity of the femur, and connected with the tibia by a strong ligament. (Fig. 13.)

The TIBIA‡ is situated at the fore and inner part of the leg. It is triangular in shape, and forms the sharp ridge which may be felt on the front part of the leg below the knee. (Fig. 1.)

The FIBULA§ is smaller than the tibia, and of similar shape. It is firmly bound to the tibia at each extremity. (Fig. 1.)

22. The BONES OF THE FOOT are the *Tar'sus* (instep), *Metatar'sus* and *Pha-lan'ges* (toe-bones). (Fig. 9.)

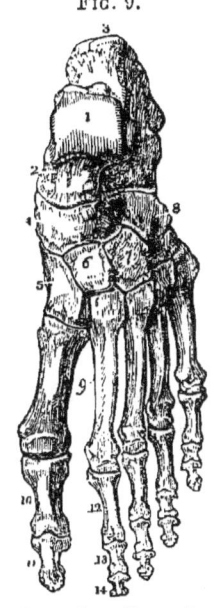

FIG. 9. THE UPPER SURFACE OF THE BONES OF THE FOOT.—1, The surface of the astragulus or ankle-bone, where it unites with the tibia. 2, The body of the astragulus. 3, Calcis or heel-bone. 4, The scaphoid. 5, 6, 7, The cuneiform. 8, The cuboid. 9, 9, 9, The metatarsal bones. 10, 11, The phalanges of the great toe. 12, 13, 14, The phalanges of the other toes.

The Sacrum. The Coccyx. Give the number and names of the bones of the Lower Extremities. Describe the Femur. Patella. Tibia. Fibula. State the names of the bones of the Foot.

* Lat., *thigh*.   † Lat., *little dish*.   ‡ Lat., *a flute*.   § Lat., *a clasp*.

The TARSUS is formed of seven irregular bones, which are so firmly bound together as to permit but little movement. (Fig. 9.)

The METATARSUS consists of five bones. The tarsal and metatarsal bones form a strong arch toward the inner and lower surface of the feet. This structure conduces to the elasticity of the step, and the weight of the body is transmitted to the ground by the spring of the arch in a manner which prevents injury to the numerous organs.

The PHALANGES* of the toes have fourteen bones; each of the small toes has three ranges of bones, while the great toe has but two. In man the great toe is not opposed to the others; in a well-formed foot the second toe is the longest. (Fig. 9.)

23. The UPPER EXTREMITIES contain sixty-four bones: the *Clav'icle* (collar-bone); the *Scap'ula* (shoulder-blade); the *Hu'merus* (arm-bone); the *Ra'dius* and *Ul'na* (fore-arm); and the bones of the *Hand*. (Fig. 1.)

The CLAVICLE,† shaped like the italic *f*, is attached at one extremity to the sternum, and at the other to the scapula. (Fig. 1.) It braces and thus prevents the shoulders from falling in toward each other.

The SCAPULA, a flat, thin, triangular bone, is situated upon the upper and back part of the chest. It lies upon muscles by which it is held in place and moved in different directions. (Fig. 30.)

The HUMERUS is a long, cylindrical bone that extends from the shoulder to the elbow. (Fig. 1.)

The ULNA ‡ is the small bone of the fore-arm, and occupies the inner side. It articulates or joins with the humerus at the elbow, forming a perfect hinge-joint. (Fig. 1.)

The RADIUS § is placed on the outside (the thumb side) of the fore-arm. It is larger than the ulna, and articulates with it, both at the elbow and at the wrist. The radius also unites

---

Describe the Tarsus. The Metatarsus. The Phalanges of the foot. State the number and give the names of the bones of the Upper Extremities. Give the form of the Clavicle. State its use. Describe the Scapula. The Humerus. What is the Ulna? With what does it articulate? Where is the Radius placed? Give its articulations.

* Gr., *row*.   † Lat., *clav'is*, a key.   ‡ It., a *measure*.   § Lat., a *spoke*.

with the first row of the bones of the hand, forming the wrist-joint.

24. The BONES OF THE HAND are the *Car'pus* (wrist); the *Metacar'pus* (palm of the hand); and *Phalan'ges* (finger-bones).

The CARPUS has eight bones, arranged in two rows, and so firmly bound together as to permit little movement of the wrist. One row articulates with the fore-arm, the other with the bones of the palm of the hand. (Fig. 1.)

The METACARPUS* has five bones, upon four of which are placed the first range of finger-bones, and upon the other the first thumb-bone. This bone of the thumb is the shortest, and it is also opposed to the other finger-bones. (Fig. 1.)

The PHALANGES of the fingers have three bones, while the thumb has but two. The fingers are named, in succession, the thumb, the index, the middle, the ring, and the little finger. (Fig. 1.)

Observation.—The wonderful adaptation of the hand to all the mechanical offices of life is one cause of man's superiority over the inferior animals. This arises from the size and strength of the thumbs and the different lengths of the fingers.

25. The JOINTS are formed by the ends of bones, usually enlarged and variously united. Generally, one surface is somewhat convex or rounded and the other concave or cup-like, the two parts being beautifully fitted to each other. All the Articulations or joints are distributed into three groups, the *Immovable*, the *Mixed* and the *Movable*.

The IMMOVABLE JOINTS have the ends of the bones placed near each other, without intervening cartilage; as, the bones of the skull and some of the bones of the face.

The MIXED JOINTS have the bones united by cartilage; as, the bones of the spinal column.

The MOVABLE JOINTS are the most perfect articulations.

Name the bones of the Hand. How are the wrist-bones arranged? Describe the bones of the palm of the hand. The finger-bones. How are Joints formed? Name the groups of articulations. Describe a Movable Joint. Give examples. What are Mixed Joints? Give an example. What is said of Movable Joints?

* Gr., *meta*, after or beyond, and *karpos*, wrist.

The bones are covered with *Cartilage,* and this surrounded by *Synovial membrane.* Outside of and connected with this membrane are the special ties, or *Ligaments,** as the joints of the upper and lower extremities. (Fig. 11.)

26. CARTILAGE is a smooth, pearly-white substance. Upon the convex surface of the bones that form a joint, the cartilage is thickest in the centre, and that which covers the concave surface is thickest around the edges. (Fig. 10.)

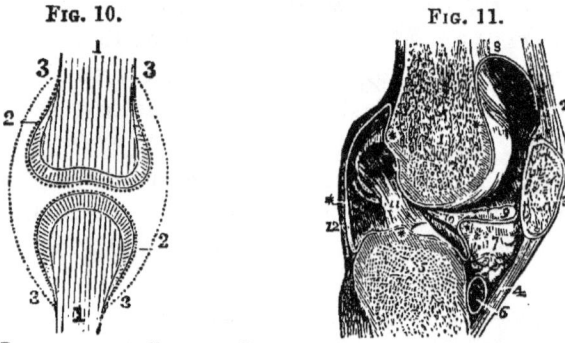

FIG. 10.   FIG. 11.

FIG. 10. A DIAGRAM OF THE RELATIVE POSITION OF THE BONE, CARTILAGE AND SYNOVIAL MEMBRANE.—1, 1, The extremities of two bones, to form a joint. 2, 2, The cartilage that covers the end of the bone. 3, 3, 3, 3, The synovial membrane, which covers the cartilage of both bones and is then doubled back from one to the other, represented by the dotted lines.

FIG. 11. A VERTICAL SECTION OF THE KNEE-JOINT.—1, The femur. 3, The patella. 5, The tibia. 2, 4, Ligaments of the patella. 6, Cartilage of the tibia. 12, The cartilage of the femur. * * * *, The synovial membrane.

The SYNOVIAL MEMBRANE is a thin, flexible skin that secretes a fluid called *Synovia* (joint-oil). It is the most perfect lubricating substance known. (Fig. 11.)

The LIGAMENTS are strong fibrous bands, more or less elastic, that bind together the joints. (Fig. 11.)

27. In all animals when very young the framework (Skeleton) is cartilaginous or animal matter; soon the pieces of cartilage become charged with a mineral substance—*lime*—by which they are made firm and somewhat brittle. This stage of development forms the bony skeleton.

Define Cartilage. What is the arrangement of Cartilage in joints? Describe Synovial Membrane. What is said of Synovia? What are Ligaments? What is said of the Skeleton of young animals?

* Lat., *ligo,* I bind.

Observation.—To show the lime without the animal matter, burn a bone in a clear fire, and it becomes white and brittle, the animal part having been consumed. To show the animal matter without the lime, immerse a slender bone for a few days in a weak acid (one part muriatic acid and six parts water), and it becomes flexible, the earthy matter having been removed.

28. The bones of the Skeleton vary in form. Some are long, others are short or broad. The long bones are hollow, or have an arched form compact upon the surface, and spongy within. This open texture increases toward the ends, which it entirely fills, excepting the very thin, hard wall. The hollow cavity is filled with a yellowish fat called *Medul'la* (marrow).

The flat bones, as those of the skull, have an outside and an inside layer of bone, with an intervening spongy texture. The bones of the spinal column and those of the wrist are less spongy than the ends of the thigh-bone, but less compact than the surface of the shaft.

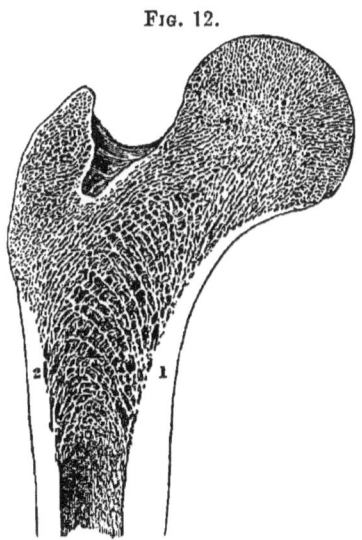

Fig. 12.

FIG. 12. LONGITUDINAL SECTION OF THE EXTREMITY OF THE FEMUR, exhibiting the arrangement of the spongy substance.— 1, 2, Positions in which the compact substance appears to resolve itself into a series of arches.

29. The bones are covered with a dense membrane called *Perios'teum.** This fibrous membrane invests the skull, and is called *Peri'cranium.*

How can the existence of animal and earthy matter be proved? Mention the different forms of bones. The structure of long bones. Of flat bones. The bones of the Spinal Column. What is Periosteum? What is this membrane called when it covers the skull-bones?

* Lat., *peri*, around, and *os*, a bone.

§ 3. PHYSIOLOGY OF THE BONES.—*Adaptation of their Structure to their Uses. General Uses of the Bones. Skill as shown in the union of the Skull with the Spinal Column. The Uses of the Joints. Classification of the Joints.*

30. THE BONES determine generally the size of the body. They support all the soft parts, as the flesh and vessels, and likewise afford a firm surface for the attachment of the liga-

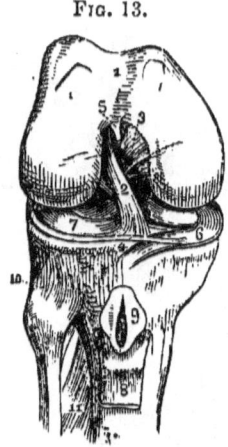

FIG. 13.

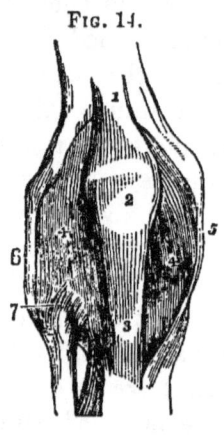

FIG. 14.

FIG. 13. THE RIGHT KNEE-JOINT, laid open from the front.—1, Articular surface of the femur. 2, 3, Ligaments. 4, Insertion of one of these ligaments into the tibia. 6, 7, Internal and external cartilages. 8, Ligament of the patella.

FIG. 14. FRONT VIEW OF THE RIGHT KNEE-JOINT.—1, Tendon of the extensor muscle. 2, Patella. 3, Ligament of the patella, or tendinous insertion of the muscle just mentioned. 4, 4, Capsular ligament. 5, 6, Internal and external lateral ligaments.

ments. In their adaptation to their several offices they exhibit a perfection of mechanism worthy the infinite mind of the Divine Architect.

The use of the various bones is different. Some protect organs, as those of the skull and chest; some for support, as the pelvis; while others are used for motion, as those of the extremities and spinal column. The bones of the upper extremities exceed all others or any instrument of art in the variety of motion and uses to which they can be put.

The union of the spinal column with the skull exhibits one of the most ingenious contrivances to be met with in the body.

Give the uses of the bones. What is said of the union of the Spinal Column with the Skull?

1st. It permits the backward and forward movement, as in bowing and nodding the head. 2d. The motion which is made in turning the head from side to side. This admirable piece of mechanism affords great protection to the spinal cord at the top of the neck, this being, perhaps, the most vital portion of the whole body. Injury to it or pressure upon it is instantly fatal.

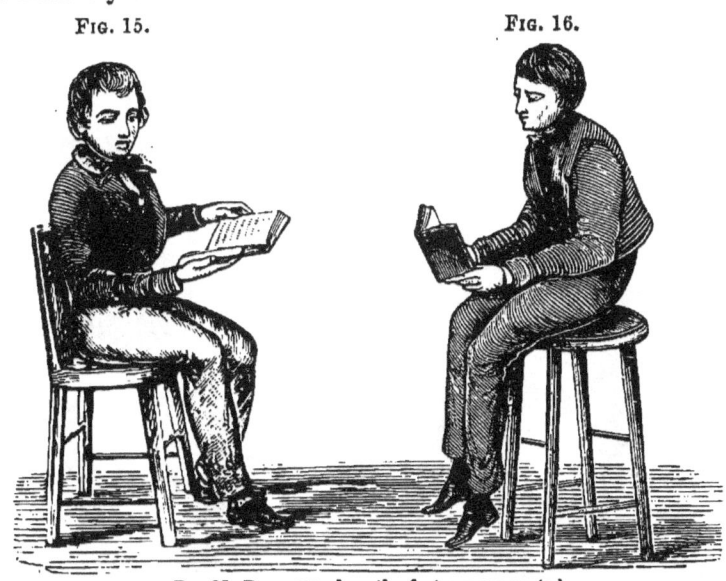

Fig. 15. Position when the feet are supported.
Fig. 16. Position when a seat is too high.

31. THE JOINTS.—The uses of the joints are to enable the body to sustain greater weight, to diminish the force of blows or shocks, to afford freedom of movement, to modify the direction in the action of muscular power.

32. For simple union without movement, we find the Immovable joint; for great strength and little movement, the Mixed joint; and for full freedom of movement, the Movable joint. Of the movable joints for motion in one plane and two directions, we find the Hinge-joint, as the knee and elbow joints; and for free rotary motion, the Ball-and-Socket joint, as the hip and shoulder joints. (Figs. 1, 6, 13.)

Enumerate the uses of the Joints. State the purposes of the different kinds of Joints.

§ 4. HYGIENE OF THE BONES.—*Effect of Exercise upon the Bones of Children. Effect of Compression—Of Stooping. Treatment of Fractures—Of Sprains—Of Felons.*

33. *The health of the bones is promoted by regular exercise.* The kind and amount of labor should be adapted to the age, health and development of the bones; neither the cartilaginous bones of the child nor the brittle bones of the aged man are

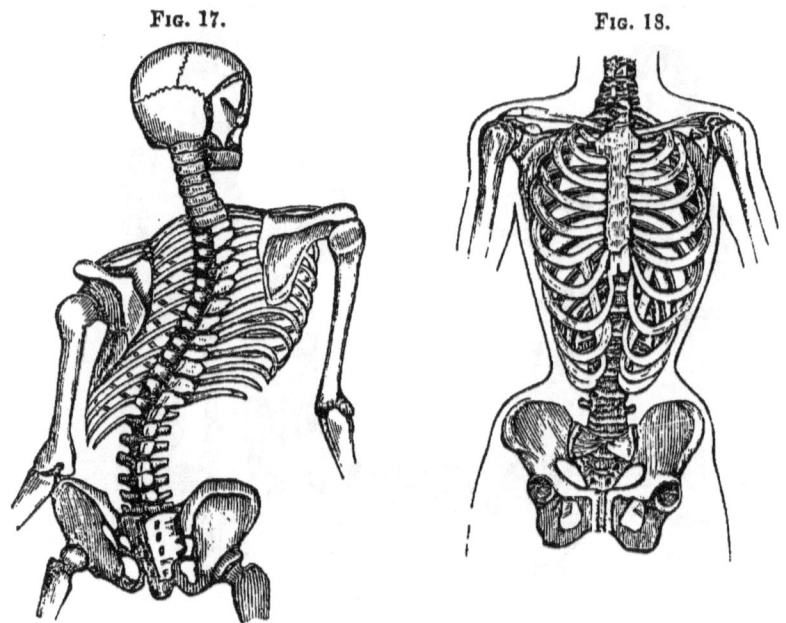

FIG. 17. A DEFORMED THORAX AND SPINAL COLUMN.
FIG. 18. A CHEST FASHIONABLY DEFORMED.

adapted to long-continued and severe exercise. While protracted exercise in childhood is injurious, moderate and regular labor favors a healthy development of the bones.

34. *The lower extremities of the very young are not adapted to sustaining much weight;* hence, to induce a child to walk, or to stand by chairs, while the bones of the lower limbs are imperfectly developed, is productive of serious injury; "bow" legs are thus produced. The benches or chairs for children

---

What is the influence of exercise on the health of the bones? To what are the lower extremities of the very young not adapted?

in a school-room should permit the feet to rest upon the floor, otherwise the weight of the limbs below the knee may cause the flexible bone of the thigh to become curved; the chairs should also have suitable backs, and the child be allowed frequent change of position. (Figs. 15, 16.)

35. *Compression of the chest should be avoided.* In youth the ribs are very flexible, and a small amount of pressure will increase their curvature, particularly at the lower part of the waist. By tight or "snug" clothing the ribs are drawn down and the space between them lessened, so that in some instances the anterior extremities of the lower ribs are brought quite together; hence, the apparel should be loose and supported by the shoulders, both for children and adults.

36. *An erect position both in sitting and standing should be carefully maintained.* The spinal column naturally curves from front to back, but not from side to side. The admirable arrangement of the bones and cartilages permits a great variety of motions and positions, the elasticity of the cartilages always tending to restore the spine to its natural position; but if a stooping or a lateral curved posture be continued for a long time, the compressed edges of the cartilages lose their power of reaction, and finally one side becomes thinned, while the other is thickened. These wedge-shaped cartilages produce permanent curvature of the spine, which is often attended with disease of the spinal cord. (Fig. 17.)

Observation 1.—The student, seamstress and artisan frequently acquire a stooping position by inclining forward to bring their books or work nearer the eyes. The desk of the pupil is often higher than the elbow as it hangs from the shoulder at rest; consequently, in drawing, writing and often in studying, one shoulder is elevated and the other depressed, distorting the spine. In the daily employments of life children should early be taught to use the left hand and shoulder more freely. Distortions of the chest necessarily accompany deformity of the spine, and disease of the heart and lungs follows, compared to which the loss of symmetry is a minor consideration.

What is said of the height of benches and chairs? Why should compression of the chest be avoided? Why should an erect position be maintained? How are distortions produced?

2.—Eminent physicians both in this country and Europe state that, among the fashionably educated, not one female in ten escapes deformities of the shoulders and spinal column. The student, to prevent as well as to cure slight curvatures of the spine, should walk with a book or a heavier weight upon the head. Porters and laborers of some countries bear very great burdens upon their heads, and walk at a rapid pace with comparative ease. Such persons, in general, have erect forms.

3.—*Fractured or diseased bones and ligaments should receive special attention.* In *fractured bones* a surgeon's care is not only needed to adjust the parts, but for several weeks to watch the reunion, that the limb may not be crooked or shortened. In *sprains* the ligaments are not usually lacerated, but strained and twisted, causing much pain, and afterward inflammation and weakness of the joints. To effect a cure there should be absolute *rest* for days, and perhaps weeks, using tepid bathing and prolonged moderate friction. More persons are crippled from ill-cared-for sprains than fractured bones. Persons enfeebled by disease, particularly scrofula, cannot be too assiduous in adopting an early and proper treatment of injured joints, to prevent the affection called "white swelling."

4.—The disease called "*Felon*" is an inflammation that commences in or beneath the periosteum. It is attended with severe, throbbing pain, and the unyielding structure of the parts prevents much swelling. The only *successful* treatment of this painful affection is an *early, free* opening through the periosteum to the surface of the bone. The earlier the incision is made, the less the risk and the suffering. The same treatment must be adopted in inflammation of large bones.

§ 5. COMPARATIVE ANATOMY (Osteology). *Classification of Animals according to their structure. General Characteristics of Vertebrates and Invertebrates. Classification of Vertebrates. Compare Bones of the Head of Vertebrates. The Vertebral Column—The Thorax—The Extremities. Characteristics of Annulosa—Mollusca—Radiata—Protozoa.*

37. IN minute structure and chemical composition all animals are essentially the same, but the different functions and habits require special conformations.

38. Animals may be separated into two divisions, *Verte-bra′ta* (Vertebrates) and *Invertebra′ta* (Invertebrates). These

---

What statement by eminent physicians? What is the prevention and cure for slight curvatures of the spine? Give observation respecting fractured bones. Sprains. Felons Name the divisions of the Animal Kingdom.

are subdivided into five sub-kingdoms,* namely, *Vert'ebrata* (Back-bone), *An'nulosa* (Ring), *Mollus'ca* (Sac), *Radia'ta* (Star), and *Protozo'a.*†

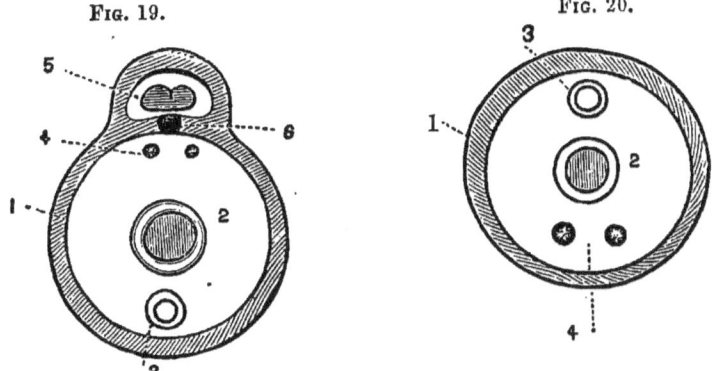

FIG. 19. DIAGRAM OF A TRANSVERSE SECTION OF A VERTEBRATE.—1, The walls. 2, The digestive organs. 3, Circulatory organs. 4, The ganglia. 5, The spinal cord. 6, Spinal column.

FIG. 20. DIAGRAM OF A TRANSVERSE SECTION OF AN INVERTEBRATE.—1, The walls. 2, The digestive organs. 3, The circulatory organs. 4, The ganglia.

39. If a VERTEBRATE is divided transversely, or cut in halves, two separate cavities are found; the upper cavity contains the main mass of the Nervous System (Brain and Spinal Cord). The lower cavity contains the Digestive and the Circulatory Systems. (Fig. 19.)

If an INVERTEBRATE is similarly divided, only one cavity is exhibited; this contains the Digestive and Circulatory Systems, with the *Gan'glia* (centres or enlargement of the nerves). (Fig. 20.)

40. The GANGLIA of Vertebrates are placed on the upper (dorsal) side of the cavity, and the circulatory organs on the lower (ventral) side. In Invertebrates the Ganglia are found

---

State the Sub-kingdoms. Give the distinctions of the two divisions. What are Ganglia? Where are they found in Vertebrates?

* The brief outlines of Zoölogy introduced in this work are arranged into *two divisions*, from Lamarck. Writers on Natural History, as Linnæus, Cuvier, Edwards, Nicholson and others, have adopted different sub-kingdoms numerically. I have chosen to arrange them into five *sub-kingdoms*.

† Gr., *protos*, first, and *Zoön*, an animal.

on the lower and the circulatory system on the upper side. (Figs. 19, 20.)

41. Vertebrates have an *internal* skeleton, generally composed of bones; some few are cartilaginous. In vertebrates the spinal column is never absent. A distinctive characteristic of vertebrates is that the brain and spinal cord are shut

Fig. 21.

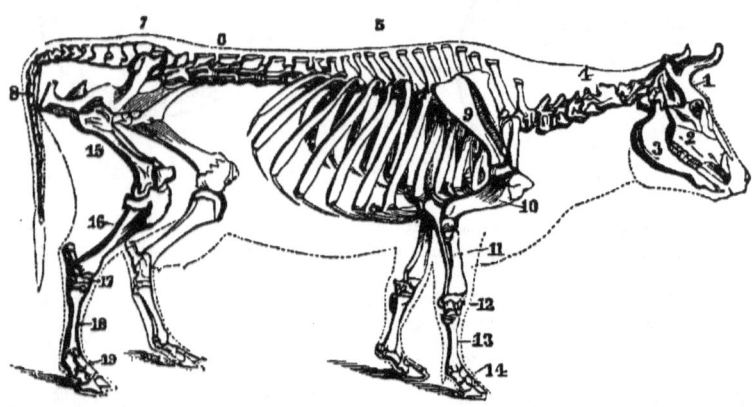

Fig. 21. Skeleton of the Cow.—1, Frontal bone of the head. 2, Upper jaw (superior maxillary). 3, Lower jaw (inferior maxillary). 4, Cervical vertebræ. 5, Dorsal vertebræ. 6, Lumbar vertebræ. 7. Sacral vertebræ. 8, Caudal vertebræ. 9, Scapula. 10, Humerus. 11, Radius and ulna. 12, Carpus. 13, Metacarpus. 14, Phalanges (toes). 15, Femur. 16, Tibia. 17, Tarsus. 18, Metatarsus. 19, Phalanges. In this figure the same terms are used as for the corresponding bones in man (see Fig. 18). The common names vary.

off from the general cavity of the body. In this sub-kingdom are classed *Mam'mals, Birds, Rep'tiles, Amphib'ians* and *Fishes*.

MAMMALS include Man and all the ordinary quadrupeds. This class are characterized by having special glands that secrete milk, by which the young are nourished after birth.

BIRDS are oviparous,* vertebrate animals, with a double circulation, and covered with feathers.

In Invertebrates? Give a characteristic of Vertebrates. State the classes of this sub-kingdom. What animals are included in the class Mammals? Name the characteristics of Birds.

* Lat., *ovum*, egg, and *pario*, to produce.

REPTILES comprise a class of vertebrates with incomplete circulation, breathe air from birth, and are generally covered with scales or plates.

AMPHIBIANS are so formed as to live on land, and for a long time under water. Their distinguishing characteristic is that they invariably undergo some kind of metamorphosis or change after birth. At first the general conformation of the body resembles fishes; at this stage they breathe by gills; subsequently they change form, and in their adult state possess air-breathing lungs. The skin is generally naked.

FISHES are oviparous, vertebrate animals, and breathe by gills. They differ in the form of the bodies, but the outline is simple. They are usually covered with scales.

42. The BONES OF THE HEAD of other *Mammals* resemble, in many points, those of man. In some quadrupeds, as the Horse and Cow, the frontal bone of the skull is in two pieces. In the Elephant the skull-bones unite in early life, and thus form but one bone. In the Hog the parietal bones of the skull are united in one bone, while the frontal bone has two pieces.

The great majority of Mammals possess teeth, which vary and constitute most important characters for separating the orders of this sub-kingdom from each other. The structure of the jaw also varies. In those animals provided with tusks there are two small bones (Intermaxillary) between the two upper jaw-bones. In the Horse, Hog and Cow the lower jaw consists of one bone. (Fig. 21.)

In *Birds* the bones of the head, in number and position, resemble Mammals, but they are early united, leaving no trace of the sutures. The upper jaw of the bird is so articulated with the skull as to admit of motion independent of the lower jaw (which never occurs in mammals), and the lower jaw, instead of being articulated directly with the skull, is connected through the intermedium of a distinct bone called the *Os Quadratum*. (Fig. 22.)

Of Reptiles. Of Amphibians. Of Fishes. What is said of the bones of the head in Mammals? Birds?

28   ANATOMY, PHYSIOLOGY AND HYGIENE.

In *Reptiles* the head-bones are irregular in form, and greatly vary in number.

In *Fishes* the bones of the head are numerous and irregular, and their study is a matter of much interest in acquiring a full knowledge of Natural History. (Fig. 24.)

43. The VERTEBRAL COLUMN of other *Mammals*, with slight modifications, is like that of man. The number of cervical

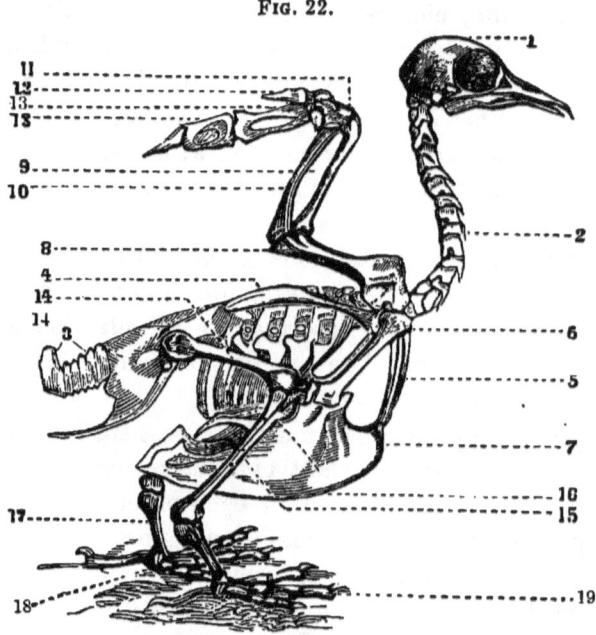

FIG. 22. SKELETON OF A BIRD.—1, The head. 2, Cervical vertebræ. 3, Dorsal and lumbar vertebræ. 4, Scapula. 5, Clavicle. 6, Coracoid bone. 7, Sternum. 8, Humerus. 9, Radius. 10, Ulna. 11, Carpus. 12, Metacarpus. 13, 13, Phalanges (fingers). 14, Femur. 15, Tibia. 16, Fibula. 17, Tarsus. 18, Metatarsus. 19, Phalanges (toes).

vertebræ is almost invariably seven; the dorsal average thirteen; the lumbar or loins, from three to seven; the sacral, usually four; the caudal or tail, from four (the number of the coccyx in man) to forty-six. The length of any part of the column seems to depend not so much upon the *number* of

---

Reptiles? Amphibians? Fishes? Compare the vertebral column of Mammals.

the vertebræ as upon their *length;* thus we find seven cervical vertebræ in the long-necked Giraffe and in the short-necked Mole.

In *Birds* the flexibility of the neck enables any part of the body to be reached by the beak. This is owing to the ball-and-socket articulations and to the great number of cervical vertebræ, which in the swan are twenty-four. The dorsal vertebræ vary from seven to eleven, and are generally consolidated into one, but in birds that do not fly they remain distinct and movable. The last caudal vertebra has a large, strong process, shaped like the letter V, for the support of the large feathers, which act as a rudder in flight.

In *Reptiles* the vertebræ vary in number from some twenty-four to four hundred, as in the Python.

In AMPHIBIANS the vertebræ may be hollow at both ends, or rounded in front and hollow behind. In Frogs the spinal column is short and the dorsal vertebræ are very long.

In *Fishes* there are but two kinds of vertebræ, the dorsal and the caudal, and these vary in number from twenty to two hundred. The vertebral bodies present a conical, cup-like depression on each side, which contains a gelatinous fluid having the same use as the elastic cartilage between the vertebræ in Mammals.

44. The STERNUM OF MAMMALS is long and narrow in shape, flat and destitute of a keel or ridge.

In *Birds* it is much extended, and forms the largest bones in their bodies. It has upon its anterior surface a *ridge* resembling the keel of a ship, for the support of the muscles of the chest used in flying. The size is proportioned to the powers of flight; hence in the little Humming-bird, which is on the wing most of the day, it reaches the maximum of development.

Of the *Reptiles*, Serpents have no sternum; but in Turtles it has an extraordinary development, and extends from the base of the neck to the commencement of the tail, forming the ventral or belly part of the shell-covering. *Fishes* have no sternum or breast-bone, properly so called.

45. The RIBS are much alike in *Mammals*, generally in twelve pairs; in the Horse, however, there are eighteen pairs.

In *Birds* the cartilage that unites the rib to the sternum is bony, giving solidity to the chest.

In some *Reptiles*, as Lizards and Crocodiles, the ribs are more numerous than in Mammals and Birds, and protect the abdomen as well as the chest. In the Turtle the ribs are expanded, forming the dorsal part of its shell, or the roof of its portable dwelling-house. In Serpents the lower or anterior extremities of the ribs have no cartilage; they aid in progressive movement or crawling, as under the skin their ends can be placed on the ground like feet.

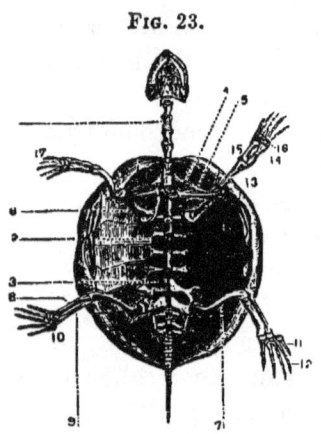

FIG. 23. SKELETON OF A TORTOISE.—1, Cervical, 2, Dorsal, 3, Lumbar vertebræ. 4, Scapula. 5, Clavicle. 6, Coracoid bone. 13, Humerus. 14, Ulna. 15, Radius. 16, Carpus. 17, Phalanges (fingers). 7, Femur. 8, Tibia. 9, Fibula. 10, Tarsus. 11, Metatarsus. 12, Phalanges (toes).

Among AMPHIBIANS Frogs and Toads have no ribs. In Newts they are rudimentary.

In some *Fishes* the ribs are wanting; in others they are very complete, and surround the trunk; in still others they are connected with a chain of bones representing the sternum.

46. The CLAVICLE maintains the shoulders apart; hence, in quadrupeds, where its presence would be a defect, it is wanting, as in the Horse and Cow.

The clavicles of *Birds* are peculiar; they unite at their anterior extremity, forming a forked bone called *fur'cula* (wish-

---

Fishes. Describe the ribs in the different classes. Why not a clavicle in the ox? Describe the clavicle of Birds.

bone). In birds of powerful flight, as the Eagle, the clavicles are very strong; in others, as the domestic Turkey, they are weak. Connecting the scapula to the sternum is the *cor'acoid* bone, which is placed side by side with the furcula, and is the main source of support to the wings in flight.

In some *Reptiles*, as the Tortoise, both the clavicle and the coracoid bone are found, while in others, as Serpents, both are wanting.

In *Fishes* the true clavicle is wanting, but in some species there is a modified form of the coracoid bone, free at its lower extremities, which may, perhaps, subserve the purpose of the clavicle of the higher animals. (Fig. 24.)

Fig. 24.

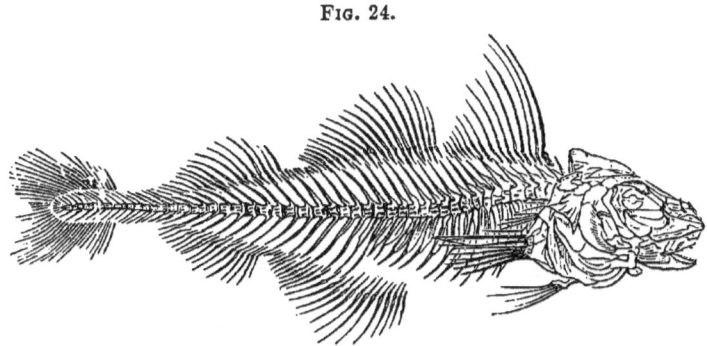

Fig. 24. The Skeleton of a Haddock.

47. The SCAPULA is present in *Mammals*, *Birds* and most *Reptiles* and *Fishes*. In the Horse and Cow it is an essential bone.

In *Birds* the scapula is long and narrow.

Reptiles, *Amphibians* and *Fishes* have, in general, the scapula, but variously modified.

48. The UPPER EXTREMITIES in *Mammals* are never wanting. In animals that swim or burrow the humerus is short, thus enabling the fore limbs to be used with force; where swiftness is required, this bone is long and slender. When

---

Reptiles. Fishes. Describe the scapula in Mammals. Speak of the upper extremities of Mammals.

the hand is used for support instead of prehension or seizing, the radius loses its power of rotation on the ulna.

The hand varies according as it is used for seizing food, swimming, flight or walking on ground more or less firm. The wrist is formed of two rows of bones; the number varies from five to eleven. The Metacarpal bones vary; in the Horse there is but one bone, called canon. The fingers are never more than five. The Cow has two. The middle finger is the most persistent, being the only one left in the Horse.

FIG. 25.

FIG. 26.

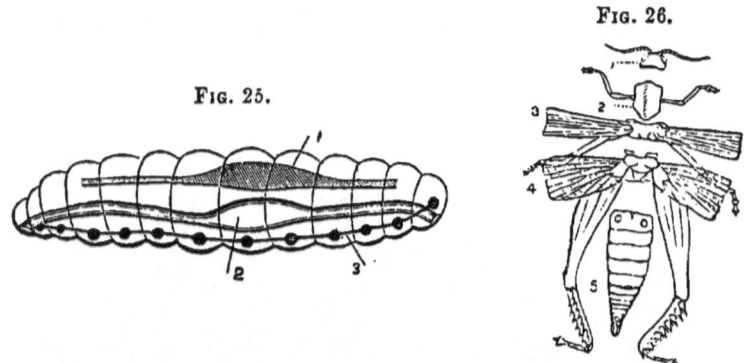

FIG. 25. DIAGRAM OF AN ANNULOSA.—1, Hæmal or vascular system. 2, Digestive organs. 3, Ganglia.

FIG. 26. DIAGRAM OF THE EXTERNAL STRUCTURE OF AN INSECT.—1, The head carrying the eyes and antennæ. 2, First segment of the thorax with the first pair of legs. 3, The second segment of the thorax with the second pair of legs and the first pair of wings. 4, The third segment of the thorax with the third pair of legs and the second pair of wings. 5, Abdomen without legs.

In the Whale the upper extremities are converted into swimming paddles.

In *Birds* the bones that form the limb proper or wing are modified to suit the special function of flight, but essentially the same parts are present as in the upper extremities of Quadrupeds.

In *Reptiles* the fore limbs are generally wanting or rudimentary.

49. The LOWER EXTREMITIES in *Mammals* are sometimes absent, as in Whales and Dolphins. In quadrupeds, as they

Birds. Reptiles. Speak of the lower extremities in Mammals.

THE BONES.   33

are used mainly for support and progression, they are less modified than the upper extremities.

In *Birds* the femur is short and straight. The tibia is the chief or longest bone of the hind limb. The fibula is united to the tibia at various distances down the leg. The ankle-joint is placed in the middle of the Tarsus.

In some *Reptiles*, as the Tortoise and Lizard, the limbs are composed of bones which in number, form, position and functions much resemble the corresponding ones in Mammals and Birds. In the Serpent tribe the limbs are wanting.

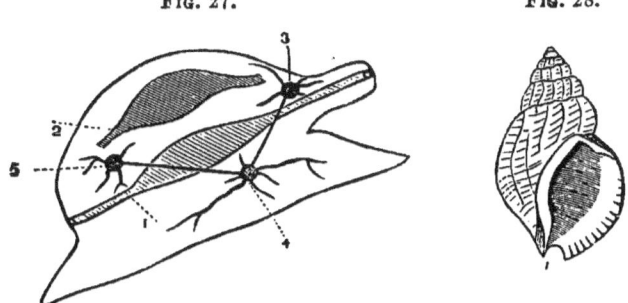

FIG. 27.   FIG. 28.

FIG. 27. DIAGRAM OF A MOLLUSCA.—1, Alimentary canal. 2, Heart. 3, Cerebral ganglion. 4, Pedal ganglion. 5, Ganglion of digestive organs and muscles.
FIG. 28. A SPECIES OF SNAIL.—1, A round month.

In *Amphibians* the limbs are well developed. In *Fishes* the extremities are rudimentary, being represented by fins.

50. The ANNULOSA* are numerous, embracing animals having an external skeleton made up of segments or rings arranged along a longitudinal line, and consisting mostly of hardened skin. This sub-kingdom is separated into two divisions, which include many classes and orders, and embrace Beetles, Weevils, Bees, Wasps, Butterflies, House-flies, Fleas, Millipedes, Centipedes, Spiders, Scorpions, Lobsters, Crabs, Worms, Leeches. (Figs. 25, 26.)

51. MOLLUSCA† are mostly soft-bodied animals that are

Birds. Reptiles. Amphibians. Fishes. Describe the sub-kingdom Annulosa. The Mollusca.

* Lat., *Annulus*, a ring.   † Lat., *Mollis*, soft.

B *

usually protected by an external skeleton or shell composed of the carbonate of lime. The Mollusca are separated into two divisions, each of which is divided into classes and orders, embracing the Nautilus, Cuttle-fishes, Snails, Limpits, Whelks, Muscles, Oysters, Scallops, Seamats, etc. (Figs. 27, 28.)

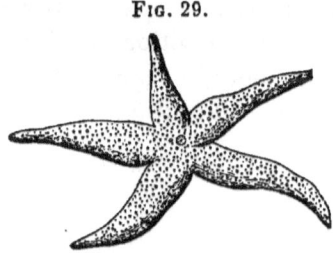

FIG. 29. DIAGRAM OF A RADIATA.—The Star-fish. 1, Mouth.

52. The RADIATA embrace animals whose organization is much less complete than that of most other animals.

53. The PROTOZOA forms the lowest division of the animal kingdom. They are generally of very minute size, and are composed of a jelly-like substance. Most of them are only seen by the aid of the microscope. They abound in the air, are largely found in water, and are popularly called "animalculæ," or "parasites." Some few attain a large size, as the sponge.

Observation 1.—The joints and limbs of domestic animals when injured or sprained should have *immediate* and *absolute rest*, particularly with the noble horse, if permanent lameness would be prevented. To allay inflammation, apply warm and even hot water, attended with rubbing or friction. If hot fomentations are not adequate after three weeks' trial, then apply a blister to the diseased part.

2.—The varied structure of the four lower sub-kingdoms of animals is replete with interest and instruction, but the necessarily limited space of this elementary school-book entirely precludes their consideration. Allow us to advise all, who can command the leisure, to extend this study to the beautiful and wonderful works of creation as seen in these parts of the garden of the Lord.

The Radiata. The Protozoa. How should injured joints be treated? Observation 2.

THE BONES.

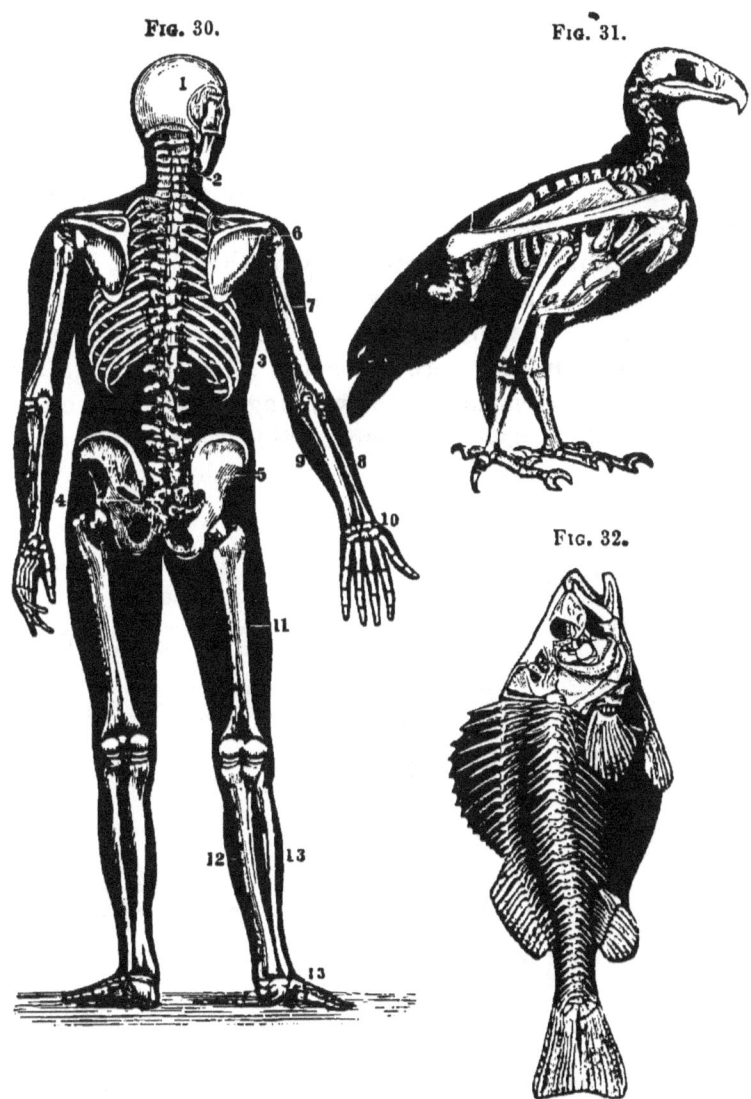

Fig. 30. Fig. 31. Fig. 32.

SYNTHETIC TOPICAL REVIEW.
OSSEOUS SYSTEM, HUMAN AND COMPARATIVE.
Classes, Sub-kingdoms, Divisions.

State the Anatomy, Physiology and Hygiene of the Osseous System, Human and Comparative, from tableaux drawn on the blackboard or outline anatomical charts.

CHAPTER III.

THE MUSCLES.

54. ALL the great motions of the body are caused by the movement of some of the bones which form the framework of the body; but these, independently of themselves, have not the power of motion, and only change their position through the action of other organs attached to them, which by contracting or shrinking draw the bones after them. In some of the slight movements, as the winking of the eye, no bones are displaced or moved. These moving, contracting organs are the *Mus'cles* (lean meat).

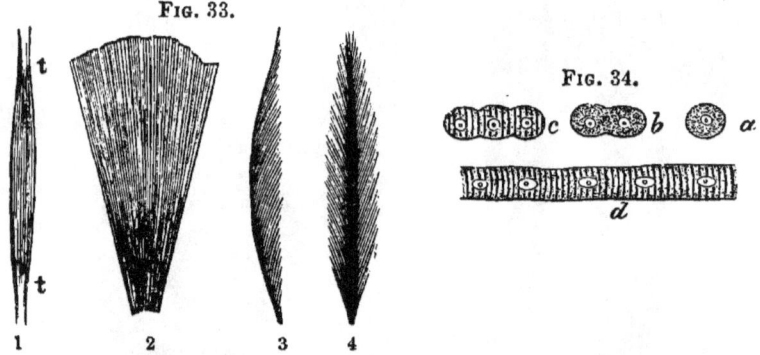

FIG. 33.—1, A REPRESENTATION OF THE DIRECTION AND ARRANGEMENT OF THE FIBRES in a spindle-shaped muscle. 2, In a radiated muscle. 3, In a peniform muscle. 4, In a bi peniform muscle. *t, t,* The tendons of a muscle.

FIG. 34. DEVELOPMENT OF MUSCULAR FIBRE FROM CELLS.—*a,* Simple cell. *b,* A pair of cells fused together. *c,* Three cells fused and their contents assuming the striated character. *d,* A muscular fibre exhibiting its original composition of cells.

§ **6.** ANATOMY OF THE MUSCLES.—*Analysis of a Muscle. Sheath of a Muscle. Tendons. Number of Muscles. Classes of Muscles. The Muscular Current.*

55. A MUSCLE is separable into bundles of fibres called *Fasciculi,* each fasciculus into *fibres,* each of the fibres into a mul-

---

How are all the great motions of the body produced? Of what are muscles composed?

titude of filaments or *fibrillæ* (fibrils), and each filament into cells arranged longitudinally.

56. Every filament, each fibre, all the fasciculi, and every muscle is surrounded by a thin, tough membrane called *Fascia* (sheath). At the extremities of the muscle the sheaths that cover the smaller fibres and the membranous covering of the whole muscle unite and form a firm inelastic cord or band called *Tendon* (sinew). (Figs. 35, 36.)

57. In some muscles the fibres run in straight lines, others spread like a fan, while others converge to one or both sides of a tendon, running the whole length of a muscle, as the plume of a feather. A few muscles that enclose cavities have the muscular fibre running in a circular direction.

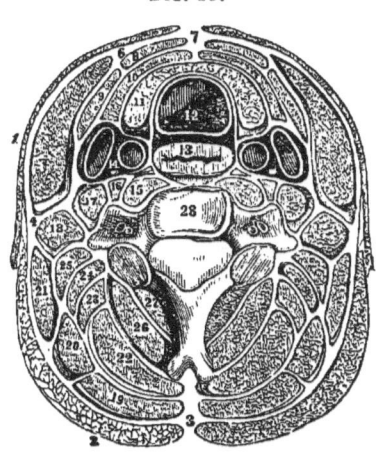

FIG. 35. A TRANSVERSE SECTION OF THE NECK. —The separate muscles, as they are arranged in layers, with their investing fascia. 12, The trachea. 13, The œsophagus. 14, Carotid artery and jugular vein. 28, One of the bones of the spinal column. (The figures in the white space represent fascia; other figures, muscles.)

58. TENDONS vary in shape. Some are long and slender, others are short and thick, while some are thin and broad. They serve to fasten the muscles to the bones or to each other.

Observation.—The different parts of a muscle, the tendon, fascia and bundle of fibres, can be seen by examining a leg of beef.

59. In general, the muscles form about the bones two layers, called the *superficial*, or external muscles, and the *deep-seated*, or those nearest the bone. There are more than four hundred muscles in the human body. To these, and a yellow

---

By what is each muscle surrounded? Describe Tendon. Give the arrangement of the fibres in different muscles. Give the different forms of Tendons. Observation. What names have been given to the layers of Muscles?

substance called *fat*, that surrounds and fills the spaces in the muscles, the child and youth are indebted for the roundness and beauty of their limbs. Muscles are classed as *Voluntary* or *Involuntary*. (Fig. 35.)

60. The VOLUNTARY Muscles are those that act when we Will or wish to use them. They are *striated*, or have beautiful parallel wavy lines, which run around the fibres in a circular direction, as the muscles of the arm and foot.

The INVOLUNTARY Muscles act independently of our wishes, and are *not striated*, or striped, as the heart. Some

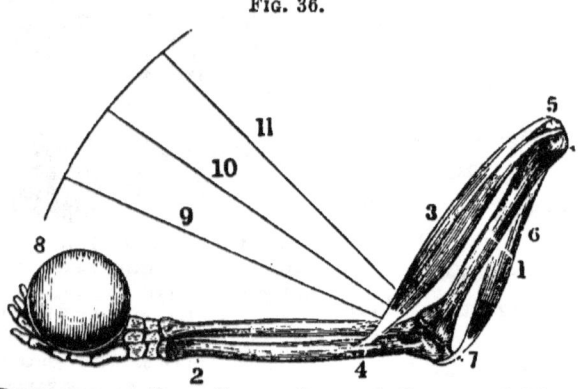

FIG. 36. DIAGRAM OF THE THIRD KIND OF LEVER.—1, Humerus. 2, Ulna. 3, Biceps muscle. 4, Its attachment to the ulna. 5, Its attachment to the humerus. 6, Triceps muscle. 7, Its tendon. 8, The ball to be moved. 9, 10, 11, Direction of the ulna and ball when the biceps (3) muscle contracts. When the triceps (6) muscle contracts, the fore-arm is extended.

muscles are both voluntary and involuntary, as the muscles used in breathing; they act with regularity when the attention is withdrawn from them, but can be controlled somewhat by our wishes. This *mixed* muscular action is highly important to the elocutionist and to the singer.

61. The chemical composition of muscles cannot be precisely known, because of the difficulty of detaching the fibres from their membranes, blood-vessels and nerves blended with them.

What is the use of fat among Muscles? How are Muscles classed? Describe Voluntary Muscles. Involuntary. Mixed Muscles. What is said of the chemical composition of Muscles?

THE MUSCLES. 39

62. In muscular action there is an increased activity in the blood-vessels in proportion to the amount of exercise performed, and also chemical changes by which heat is produced. The electrical current known as the "*muscular current*," is probably a result of chemical action. In the entire muscle its path lies along the outside toward the tendons. The direction of the total current of the body is from the head downward.

Observation.—In friction, or rubbing the body with the hand, the direction of the current should be followed; otherwise, irritation is produced rather than the soothing influence desired. This direction is of special importance to nurses and watchers in caring for the sick, particularly nervous patients. The effect of friction is sometimes improved by moistening the inside of the hand.

§ 7. PHYSIOLOGY OF THE MUSCLES.—*Relative Uses of the Bones and Muscles. Characteristic Property of Muscles. Uses of Tendons.*

63. To give a clear idea of the relative uses of the Muscles and Bones, we quote the comparison of another: "The Bones

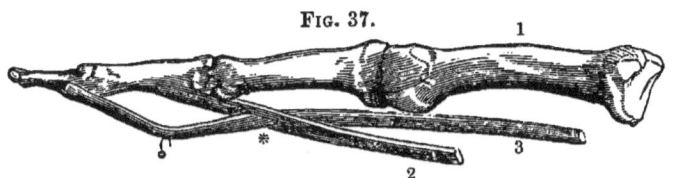

FIG. 37.

FIG. 37. METACARPAL AND PHALANGEAL BONES OF THE FINGERS, WITH THEIR TENDONS AND LIGAMENTS.—1, Metacarpal bone. 2, Tendon of the superficial flexor. 3, Tendon of the deep flexor passing through a perforation (*) of the superficial flexor.

are to the body what the masts and spars are to the ship—they give support and the power of resistance; the Muscles are to the bones what ropes are to the masts and spars."

64. The characteristic property of the muscles is *contractility*. Each fibre of the several muscles receives from the brain, through the nervous filament appropriated to it, a certain influence called *nervous fluid*, or *stimulus*. It is this that induces contraction, while the suspension of this stimulus causes

---

Name some of the changes attending muscular action. What is said of the muscular current? Observation. State the relative uses of Bones and Muscles. What is the characteristic property of Muscles? What induces contraction of Muscles?

relaxation of the fibres. During contraction the muscle shortens in the direction of its length, and enlarges in the direction of its thickness.

65. Muscles remain contracted but a short time; then they relax or lengthen, which is their rest. When the muscles are in a state of contraction they are full, hard, and more prominent than when relaxed. The muscles passing over to straighten a joint are usually called *Extensors*, because they serve to extend the part beyond the joint, while those lying in front of the joint are, for the opposite reason, called *Flexors*. (Figs. 38, 39.)

Observation.—The alternate contraction and relaxation of the muscles may be shown by clasping the fore-arm about three inches below the elbow, then open and shut the fingers rapidly, and the swelling and relaxation of the muscles on the opposite sides of the arms, alternately with each other, will be felt corresponding with the movement of the fingers. While the fingers are bending, the inside muscles swell and the outside ones become flaccid, and while the fingers are extending, the inside muscles relax and the outside ones swell. The alternate swelling and relaxation of opposing muscles may be felt in all the movements of the limbs.

66. TENDONS serve to convey the contractile power of muscles to the bones; they are in themselves passive organs, possessing no contractility. In them the evidence of care and skillful arrangement is beautifully exhibited. Wherever muscular action is wanted and the presence of muscle would be inconvenient or mar the harmony of proportion, or where great strength is needed, there we find the small, dense, conducting tendons; the slits in the short tendons of the second joint to allow the long tendons from the muscles of the forearm to pass through to the last bones of the fingers afford the best conceivable arrangement for compactness, delicacy, beauty and utility. (Fig. 37.)

What the relaxation of muscular fibre? When are Muscles called Extensors? Flexors? Observation. State the office of Tendons.

THE MUSCLES. 41

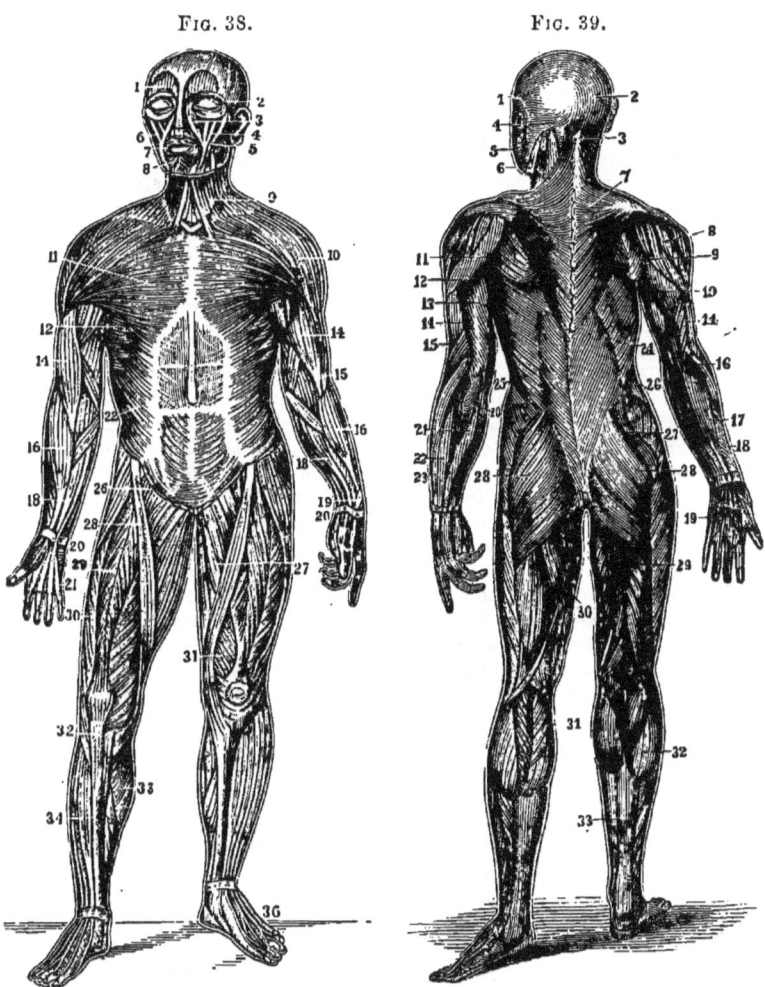

FIG. 38. A FRONT VIEW OF THE SUPERFICIAL MUSCLES OF THE BODY.—1, The frontal swells of the occipito-frontalis. 2, The orbicularis palpebrarum. 3, The levator labii superioris alæque nasi. 4, The zygomaticus major. 5, The zygomaticus minor. 6, The masseter. 7, The orbicularis oris. 8, The depressor labii inferioris. 9, The latysma myoides. 10, The deltoid. 11, The pectoralis major. 12, The latissimus dorsi. 13, The serratus major anticus. 14, The biceps flexor cubiti. 15, The triceps extensor cubiti. 16, The supinator radii longus. 17, The pronator radii teres. 18, The extensor carpi radialis longior. 19, The extensor ossis metacarpi pollicis. 20, The annular ligament. 21, The palmar fascia. 22, The obliquus externus abdominis. 26, The psoas magnus. 27, The abductor longus. 28, The sartorius. 29, The rectus femoris. 30, The vastus externus. 31, The

Name the Muscles from Figs. 38 and 39.

vastus internus. 32, The tendo patellæ. 33, The gastrocnemius. 34, The tibialis anticus. 35, The tibia. 36, The tendons of the extensor communis.

FIG. 39. A BACK VIEW OF THE SUPERFICIAL MUSCLES OF THE BODY.—1, The temporalis. 2, The occipito frontalis. 3, The complexus. 4, The splenius. 5, The masseter. 6, The sterno-cleido mastoideus. 7, The trapezius. 8, The deltoid. 9, The infra spinatus. 10, The triceps extensor. 11, The teres minor. 12, The teres major. 13, The tendinous portion of the triceps. 14, The anterior edge of the triceps. 15, The supinator radii longus. 16, The pronator radii teres. 17, The extensor communis digitorum. 18, The extensor ossis metacarpi pollicis. 19, The extensor communis digitorum tendons. 29, The olecranon and insertion of the triceps. 21, The extensor carpi ulnaris. 22, The auricularis. 23, The extensor communis. 24, The latissimus dorsi. 25, Its tendinous origin. 26, The obliquus externus. 27, The gluteus medius. 28. The gluteus magnus. 29, The biceps flexor cruris. 30, The semi-tendinosus. 31, 32, The gastrocnemius. 33, The tendo-Achilles.

Practical Explanation.—The muscle 1, Fig. 38, by its contraction, raises the eyebrows. The muscle 2, Fig. 38, closes the eyelids. The muscle 3, Fig. 38, elevates the upper lip. The muscles 4, 5, Fig. 38, elevate the angles of the mouth. The muscles 6, Fig. 38, and 5, Fig. 39, bring the teeth together. The muscle 7, Fig. 38, closes the mouth. The muscle 8, Fig. 38, depresses the lower lip. The muscles 9, Fig. 38, and 6, Fig. 39, bend the neck forward. The muscles 3, 4, Fig. 39, elevate the head and chin. The muscle 22, Fig. 38, bends the body forward and draws the ribs downward. The muscle 11, Fig. 38, brings the shoulder forward. The muscle 7, Fig. 39, draws the shoulder back. The muscles 10, Fig. 38, and 8, Fig. 39, elevate the arm. The muscles 11, Fig. 38, and 24, Fig. 39, bring the arm to the side. The muscle 14, Fig. 38, bends the arm at the elbow. The muscle 10, Fig. 39, extends the arm at the elbow. The muscles 16, 18, Fig. 38, bend the wrist and fingers. The muscle 19 bends the fingers. The muscles 18, 21, 23, Fig. 39, extend the wrist. The muscle 23, Fig. 39, extends the fingers. The muscles 26, 27, 28, Fig. 39, bend the lower limbs on the body at the hip. The muscle 28, Fig. 38, draws one leg over the other (the position of a tailor when sewing). The muscles 27, 28, Fig. 38, extend the lower limbs of the body at the hip. The muscles 29, 30, 31, Fig. 38, extend the leg at the knee. The muscles 29, 30, Fig. 39, bend the leg at the knee. The muscles 34, 36, Fig. 38, bend the foot at the ankle and extend the toes. The muscles 31, 32, 33, Fig. 39, extend the foot at the ankle.

§ 8. HYGIENE OF THE MUSCLES.—*An Essential Muscular Law. Importance of using Muscles in Pure Air—In Light. Of Exercise. Conditions to be observed in Muscular Exercise. Education of Muscles. Proper Muscular Tension.*

67. **The muscles should be abundantly supplied with pure blood** is the first and essential law. A pure state of the blood requires that the digestive apparatus should be in a healthy condition; that the vital organs should have ample volume; that the lungs should be plentifully supplied with pure air; that the skin should be kept warm by proper clothing and clean by bathing, and that it should be acted upon by *air* and *sunlight*.

Give the practical explanation of Figs. 38, 39. What is essential in possessing healthy muscles?

THE MUSCLES. 43

68. *The muscles should be used in pure air.* The purer the air we breathe, the longer can the muscles be used in labor, walking or sitting, without fatigue and injury; hence the benefit derived in thoroughly ventilating all inhabited rooms.

Observation.—It is a common remark that sick persons will sit up longer when riding in a carriage than in an easy chair in the room where they have lain sick. In the one instance they breathe pure air; in the other, usually, a confined, impure air.

69. *The muscles should be exercised in the light.* Light, particularly that of the sun, exercises as great an influence on man as it does on plants. Both require the stimulus of this agent. Students should take their exercise during the day rather than in the evening, and the farmer and the mechanic should avoid night toil, as it is much more exhausting than the same effort during daylight.

Illustration.—Plants that grow in the shade are of lighter color and more feeble than those that are exposed to sunlight. Persons that dwell in dark rooms are paler and less vigorous than those who inhabit apartments well lighted and exposed to solar light.

70. *The muscles should be used and then rested.* When the muscles are exercised, the flow of blood in the arteries and veins is increased, hence the muscular fibre increases in size and acts with greater force; while, on the contrary, the muscle that is little used receives little nutriment from the sluggish blood, and decreases in size and power.

Illustration.—The muscles of the blacksmith increase in size and become firm and hard; those of the student, if not used in gymnastics or otherwise, decrease in size and become soft and less firm.

71. *Exercise should be regular and frequent.* The system needs this means of invigoration as regularly as it does new supplies of food. It is no more correct that we devote several days to a *proper* action of the muscles and then spend one day inactively, than it is to take a *proper* amount of food for several days and then for a season withdraw this supply.

72. *Every part of the muscular system should have its appropriate share of exercise.* Some employments call into exercise

---

Why should we work in pure air? In light? How does exercise promote the health of muscles? Illustration.

the muscles of the upper limbs, as shoemaking; others the muscles of the lower limbs; while some the muscles of both upper and lower limbs, with those of the trunk, as farming.

Fig. 40. Fig. 41.

Fig. 40.—1, A perpendicular line from the centre of the feet to the upper extremity of the spinal column where the head rests. 2, 2, 2, The spinal column with its three natural curves. Here the head and body are balanced upon the spinal column and joints of the lower extremities, so that the muscles are not kept in a state of tension. This erect position of the body and head is always accompanied with straight lower limbs.

Fig. 41.—1, A perpendicular line from the centre of the feet. 2, Represents the unnatural curved spinal column and its relative position to the perpendicular (1). The lower limbs are seen curved at the knee, and the body is stooping forward. While standing in this position the muscles of the lower limbs and back are in continued tension, which exhausts and weakens them.

Those trades and kinds of exercise are most salutary in which all the muscles have their due proportion of action, as this tends to develop and strengthen them equally.

What kind of exercise is most salutary to muscular action?

73. *The proper time for exercise should be observed.* As a general rule, the morning is a better time for exercise than the evening; the powers of the system are greatest at that time. Severe exercise should be avoided immediately before or after a meal; the vigor of the system is then required for the digestive functions. The same rule should be observed regarding mental toil, as the powers of the system are then concentrated upon the brain.

74. *The erect attitude lessens the exhaustion of the muscles.* A person will stand longer, walk farther and do more work when erect than in a stooping posture, because the muscles of the back, in stooping, are in a state of tension, or stretching, to keep the head and trunk from falling forward. In the erect position the head and trunk are nicely balanced and supported by the bones of the spinal column, and the muscles of the back are called but slightly into action.

Observation.—The attitude of children in standing has been much neglected both by parents and teachers. Let a child acquire the habit of inclining his head and shoulders, and the chest will become contracted, the muscles of the back enfeebled, and the deformity thus acquired will progress to advanced age.

75. *Muscles should be rested gradually when they have been vigorously used.* If a person has been making great muscular exertion, instead of sitting down to rest, he should continue muscular action by some moderate labor or amusement.

Observation 1.—When the skin is covered with perspiration (sweat) from muscular action, avoid sitting down " to cool " in a current of air; rather put on more clothing and continue to exercise moderately.

2.—In cases when severe action of the muscles has been endured, bathing and rubbing the skin over the joints that have been used are of much importance. This will prevent soreness of the muscles and stiffness of the joints.

76. *A slight relaxation of the muscles tends to prevent their exhaustion.* In walking, dancing and learning to write there will be less fatigue and the movements will be more graceful

---

State the proper time for exercise. Why do the Muscles require erect position of the body? What attention should be given to children and youth? Give the treatment and illustrations of Muscles that have been vigorously used.

when the muscles are slightly relaxed than when rigidly contracted. The same principle applies to most of the mechanical employments.

Observation 1.—When riding in cars and coaches the system will not suffer so severely from the jar if the muscles are slightly relaxed. When riding over uneven places in roads, rising slightly upon the feet diminishes the shock occasioned by the sudden motion of the carriage. The muscles, under such circumstances, are to the body what elastic springs are to a carriage.

2.—In jumping or falling from a carriage or any height the shock to the organs of the body may be obviated in the three following ways:

Fig. 42.

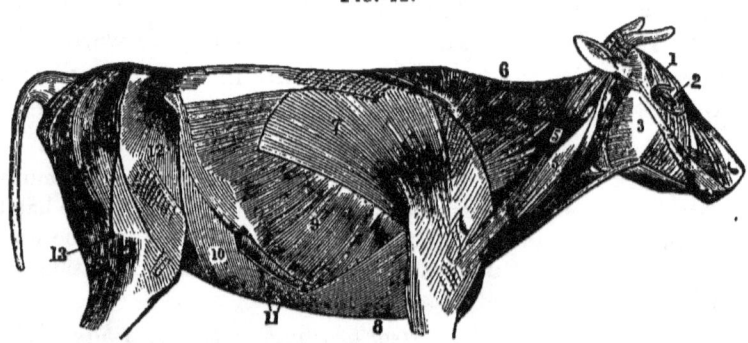

FIG. 42. SUPERFICIAL MUSCLES OF A COW.—1, Occipito-Frontalis. 2, Orbicularis Palpæbrarum. 3, Masseter. 5, Sterno-cleido-Mastoid. 6, Trapezius. 7, Latissimus Dorsi. 8, Pectoralis. 9, 10, External and Internal oblique muscle. 11, Opening of the mammary artery and vein (milk-veins). 12, Glutei. 13, Rectus Femoris muscle.

1st. Let the muscles be relaxed, not rigid. 2d. Let the limbs be bent at the ankle, knee and hips; the head should be thrown slightly forward, with the trunk a little stooping. 3d. Fall upon the toes, not the heel.

77. *Repetition of muscular action is necessary.* To render the action of the muscles complete and effective they must be called into action repeatedly and at proper intervals. This education must be continued until not only each muscle, but every fibre of the muscle, is fully under the control of the Will. In this way persons become expert penmen, singers, and skillful in every employment.

Observation.—It is exceedingly important that correct movements be insisted upon at the *commencement* of any muscular training, as it is very

---

Why should muscular action be repeated? Observation.

difficult to change a movement which has been long practiced. If a child holds his pen improperly during his early lessons, he will probably never become an easy and elegant writer.

78. *The state of the mind affects muscular contraction.* A person who is cheerful and happy will do more work and with less fatigue than one who is peevish and discontented.

Illustration.—A sportsman will pursue his game miles without fatigue, while his attendant, not having any mental stimulus, will become weary.

79. *Relaxation must follow contraction, or rest must follow exercise.* Exercise too long continued produces exhaustion, and in the exercise of exhausted muscle the loss of material exceeds the deposit; also long-continued tension enfeebles, and at length destroys, the contractile property.

Illustration.—The effect of over-work may be seen in the attenuated frames of over-tasked domestic animals, as the horse, or in the diminished weight of the farmer after the hurry of harvest-time. The effect of continued tension is seen in the restlessness of children at school after sitting for a time in one position. The necessity of frequent recesses is founded upon the organic law that relaxation of muscle must follow contraction. The younger and feebler the pupils the greater is this necessity.

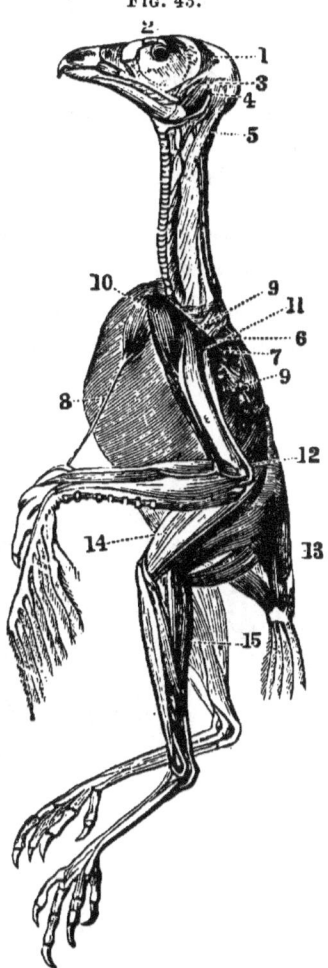

FIG. 43. SUPERFICIAL MUSCLES OF A HAWK.—1, Occipito-Frontalis. 2, Orbicularis Palpæbrarum. 3, Temporal. 4, Masseter. 5, Sterno-cleido-Mastoid. 6, Trapezius. 7, Latissimus Dorsi. 8, Pectoralis. 9, Deltoid. 10, Biceps. 11, Triceps. 12, Glutei. 13, Levator Caudæ. 14, Rectus Femoris. 15, Gastrocnemius muscle.

80. *Change of employment often affords the required rest,* as

---

State the effect of the mind on muscular contraction. The relation between relaxation and contraction of muscles. Illustration. What effect has change of employment?

it brings into action a new set of muscles; hence, the person of sedentary occupation is rested by general muscular exercise, while the person of active occupation is rested by that of a sedentary character.

Illustration.—The needlewoman exhausts the muscles of the back and arm; a brisk walk or some active household employment affords rest.

§ **9.** COMPARATIVE MYOLOGY.—*Compare Muscles of other Mammals with those of Man. Muscles of Birds—Of Reptiles—Of Fishes.*

81. In all *Mammals* the Muscles in their general plan resemble those of Man, the modifications in number, form, po-

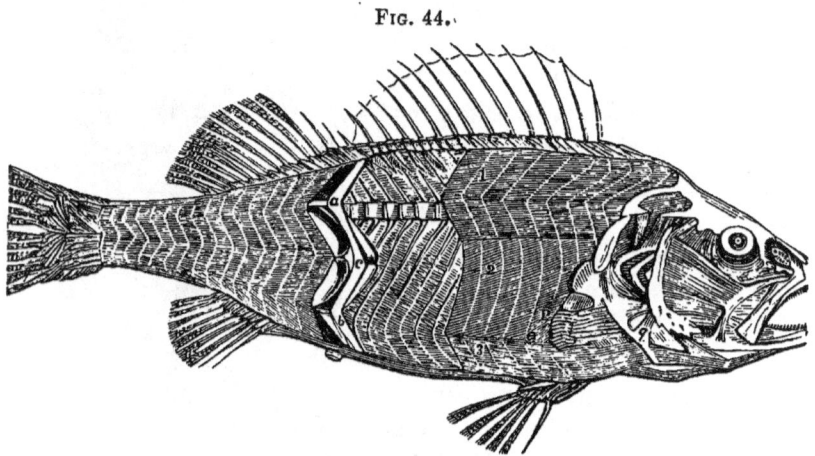

FIG. 44.

FIG. 44. MUSCLES OF THE FISH.—*a, b, c,* and 1, 2, 3, The zigzag arrangement of the flakes.

sition and relative size being only such as adapt them to the habits and necessities of the particular species. The color of the muscle is deepest in the Carnivora (flesh-eaters), and palest in the Rodentia (gnawers).

82. In *Birds* the muscular system is remarkable for the distinctness and density of their fibres, for the deep-red color of those employed in vigorous action and their marked line of attachment to their tendons, which are of a pearly-white color and have a peculiar tendency to become bone.

What is said of the muscles of Mammals? Of their color? For what is the muscular system of Birds remarkable?

83. In *Reptiles* the muscles are pale in color, and the fibres are tenacious of their contractility; the energy of their contraction in some instances and on some occasions is great, but it cannot be continuously exercised. The muscles of the trunk reach the highest development in Serpents and the lowest in the Tortoise.

84. In *Fishes* the muscular tissue is usually colorless; sometimes it is opaline or yellowish, but it is white when boiled. The chief masses of the muscular system are disposed on each side of the trunk in a series of *plates* or *flakes* corresponding in number to the vertebræ. Each flake is arranged in a zigzag manner.

Observation 1.—In a very large portion of the four lower sub-kingdoms of animals (composing the Invertebrates) the muscles are so minute that they cannot be so well demonstrated, yet it is probably true that in structure and in varied use they may be as complete and wonderful as those of the Vertebrates.

2.—Among all domestic animals, as the horse, ox and fowl, the same hygienic laws relative to protection, food, air, light and exercise are equally applicable as to man.

Speak of the muscles of Reptiles. Speak of the muscles in Fishes. What is said of the minuteness of the muscles in some animals? What of the application of hygienic laws?

SYNTHETIC TOPICAL REVIEW.
MUSCULAR SYSTEM, HUMAN AND COMPARATIVE.
Classes, Sub-kingdoms, Divisions, Anatomy, Physiology and Hygiene.

FIG. 45.

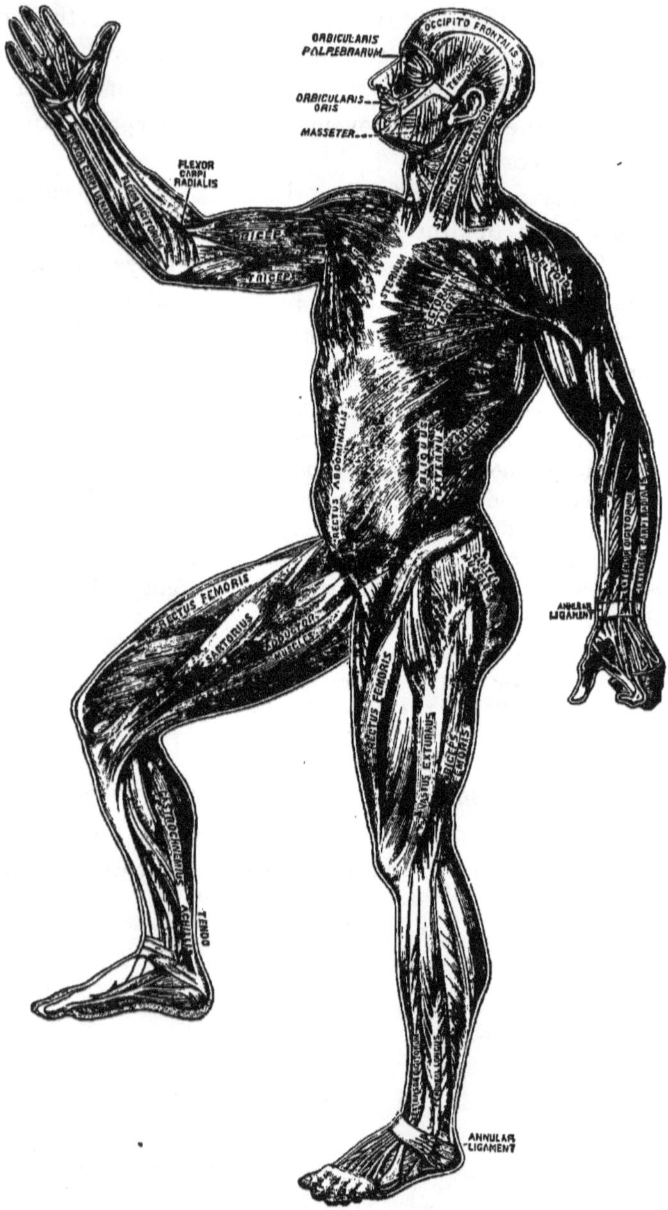

State the Anatomy, the Physiology, the Hygiene of the Muscular System, Human and Comparative, from Tableaux drawn on blackboard or from outline Anatomical charts.

DIVISION II.

THE NUTRITIVE APPARATUS.

85. IN the building up and repairing of the system, from the earliest period of embryo life to the last moment of earthly existence, certain organs are used, which together may be termed the NUTRITIVE APPARATUS, including the *Digestive*, the *Absorptive*, the *Respiratory*, the *Circulatory*, and the *Assimilatory* organs.

CHAPTER IV.

THE DIGESTIVE ORGANS.

86. THE food, whether animal or vegetable, has no resemblance to the bones, muscles and other parts of the body to which it gives sustenance. It must undergo certain essential alterations before it can become a part of the different structures of the body. The first change is effected by the action of the *Digestive Organs.*

§ **10.** ANATOMY OF THE DIGESTIVE ORGANS.—*Anatomy of the Mouth—The Teeth—The Salivary Glands—The Pharynx—The Œsophagus—The Stomach—The Intestines—The Liver--The Pancreas—The Spleen.*

87. The DIGESTIVE ORGANS include the *Mouth, Teeth, Sal'ivary Glands, Palate, Pharynx, Œsoph'agus, Stomach, Intestines, Liver, Pan'creas* and *Spleen.*

The MOUTH is the space bounded by the lips in front, the soft palate behind, the hard palate above and the floor below, upon which rests the tongue. (Fig. 46.)

The TEETH are attached to the upper and the lower jawbone by means of bony sockets, called *alve'olar* processes.

Name the organs of the Nutritive Apparatus. In what organs is the first change of food effected to make blood, bone and muscle? Enumerate the Digestive Organs. Describe the Mouth. What is said of the Teeth?

The attachment is strengthened by the fibrous, fleshy structure of the gums. Each tooth has two parts, the *crown* and the *root*. The crown is that part which protrudes from the jaw-bone and gum, and is covered by the enamel; the root or fang is that part contained in the socket of the jaw, and the slightly-constricted portion clasped by the gums is the *neck*.

Fig. 46.

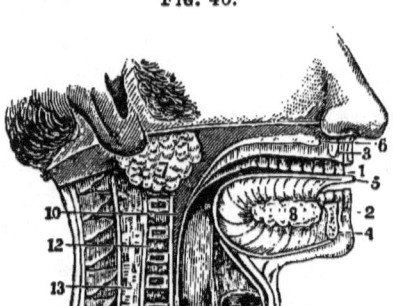

Fig. 46. The Mouth and Neck Laid Open.— 1, The teeth. 3, 4, Upper and lower jaws. 5, The tongue. 7, Parotid gland. 8, Sublingual gland. 9, Trachea (wind-pipe). 10, 11, Œsophagus (gullet). 12, Spinal column. 13, Spinal cord.

Observation.—The first set of teeth appearing in infancy is called *temporary*, or the milk teeth. They are twenty in number, ten in each jaw. Between six and fourteen years of age they are replaced by the second set, called *permanent* teeth, numbering thirty-two, sixteen in each jaw. The four front teeth in each jaw are called *Incisors* (cutting teeth). The next tooth on each side the *Cuspid* (eye-tooth in the upper jaw and stomach-tooth in the lower), the next two *Bi-cuspids* (small grinders), the next two *Molars* (grinders), situated behind the other teeth. The last molars are the *dens sapientiæ* or "wisdom teeth," smaller than their fellows, late in their development and early in their decay. The incisors, cuspids and bi-cuspids have each but one root; the molars of the upper jaw have three roots, those of the lower jaw two roots. (Fig. 47.)

88. Glands are either small sacs with openings more or less contracted, or minute tubes. Glands receive both blood-vessels and nerves. The *Salivary Glands* (glands of the mouth) consist of three pairs, the *Parot'id*,* the *Submax'illary*† and the *Subling'ual*.‡

Give the parts of each tooth. What are the temporary teeth? The permanent? Name and describe the different forms of the teeth. Define Glands. What do they receive? Name the glands of the mouth.

* Gr., *para*, near, and *ous*, ear. † Lat., *sub*, under, *maxilla*, jaw-bone.
‡ Lat., *sub*, under, and *lingua*, the tongue.

THE DIGESTIVE ORGANS.

The PAROTID GLAND, the largest, is situated in front of the external ear and behind the angle of the jaw.

The SUBMAXILLARY GLAND is situated within the lower jaw anterior to its angle.

The SUBLINGUAL GLAND is elongated and flattened, and situated beneath the floor of the mouth. Ducts from these glands open into the mouth. (Figs. 46, 48.)

Observation.—The "mumps" is a disease of the parotid gland, and the swelling under the tongue called the "frog" a disease of the sublingual gland.

FIG. 47.

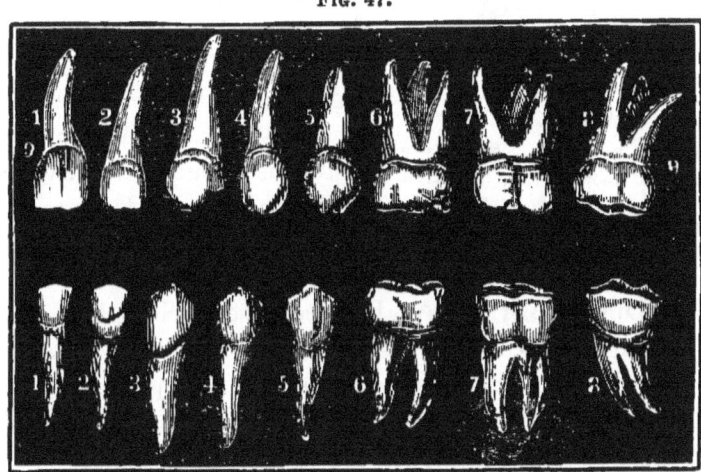

FIG. 47 REPRESENTS THE ADULT TEETH.—1, 2, The Cutting teeth (incisors). 3, Eye-tooth (cuspid). 4, 5, Small grinders (bi-cuspids). 6, 7, 8, Grinders (molars). 9, 9, Neck of the tooth.

89. The PHARYNX (throat) is the funnel-like cavity immediately following the mouth. It receives food from the mouth, and the air in breathing passes by the same passage when the nostrils are closed. Several passages lead from this cavity. (Fig. 48.)

90. The ŒSOPHAGUS (food passage) is a large membranous tube, extending from the pharynx to the stomach. It lies

---

Describe each pair of glands. Give observation. Describe the Pharynx. What is the Œsophagus?

behind the trachea (wind-pipe), the heart and the lungs, and passes through the diaphragm or floor to the chest. (Fig. 46.)

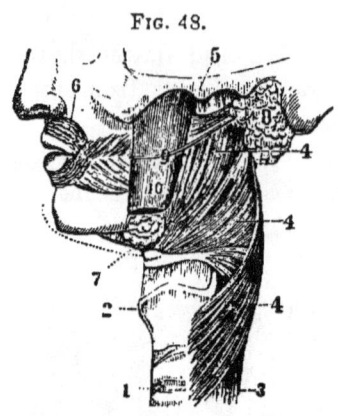

FIG. 48. A SIDE VIEW OF FACE.—1, 2, Trachea. 3, Œsophagus. 7, Submaxillary. 8, Parotid gland. 9, Duct from the Parotid gland. 4, 4, 4, 5, 6, Muscles.

91. The STOMACH is pear-shaped. When moderately filled it is about nine inches in length by three inches in diameter. It has two openings; one, connected with the œsophagus, called the *car'diac** orifice; the other, connected with the upper portion of the small intestine, called the *pylor'ic*† orifice. (Fig. 51.)

92. The INTESTINES are divided into the *Small* and the *Large* intestines. The small intestine is about twenty-five feet in length, and divided into three parts, the *Duode'num*, the *Jeju'num* and the *Il'eum*.

The DUODENUM is so called because its length is about twelve fingers' breadth, or ten inches.

JEJUNUM signifies *fasting*, the food passing quickly through this portion, leaving it empty.

ILEUM, *twisted*, is so named from its numerous coils.

93. The LARGE INTESTINE, about five feet in length, is also divided into three parts, the *Cœ'cum*, the *Colon* and the *Rectum*. (Fig. 49.)

The CŒCUM is so called from its forming a *blind pouch* (shut sac), open at one end.

The COLON, because the food passes slowly through its folds, and the RECTUM, from its *straight* course.

Attached to the extremity of the cœcum is a worm-shaped

What is said of the Stomach? Of its openings? Give the divisions of the Intestines. Name and describe the divisions of the small intestines. State the length and parts of the large intestines. Describe each part.

* Gr., *kardia*, heart. † Gr., *pulōrus*, gatekeeper.

tube (*appendix vermiformis*), about four inches long and the size of a goose-quill. Its function is unknown.

The lower portion of the colon makes a double curvature called the *sig'moid flex'ure*. The rectum extends from the sigmoid flexure to the terminus of the intestinal canal, a distance of six or eight inches. (Fig. 49.)

94. The LIVER is the largest glandular organ in the body, weighing about four pounds. It is situated in the right side below the diaphragm. It is convex above and slightly concave below. On the under side of the liver is the gall-bladder, or reservoir for the bile, which opens by a duct into the small intestine. (Fig. 112.)

95. The PANCREAS* is a long, flattened organ, having no fat, and weighing three or four ounces. It is placed behind the stomach. A duct from this organ opens into the small intestine near the bile duct. (Fig. 112.)

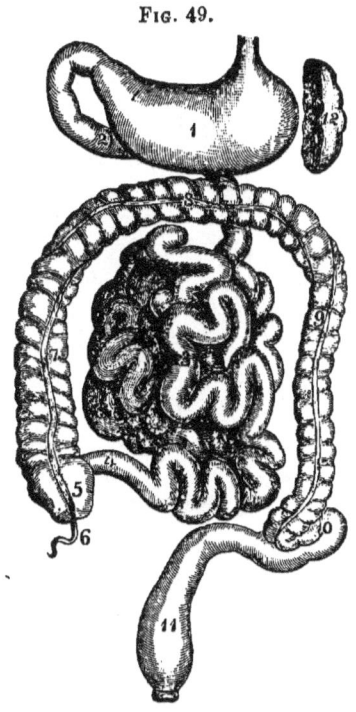

FIG. 49. THE STOMACH AND INTESTINES.— 1, Stomach. 2, Duodenum. 3, Small intestine. 4, Termination of the ileum. 5, Cœcum. 6, Vermiform appendix. 7, Ascending colon. 8, Transverse colon. 9, Descending colon. 10, Sigmoid flexure of the colon. 11, Rectum. 12, Spleen.

96. The SPLEEN (milt), so called because the ancients supposed it to be the seat of melancholy, is an oblong, flattened organ, situated on the left side in contact with the diaphragm, stomach and pancreas. It is of a dark-bluish color, has no outlet, and its use is not well determined. (Fig. 49.)

Describe the Liver. Describe the Gall-bladder. What is the shape of the Pancreas? What peculiarity? What is said of the Spleen?

* Gr., *pan*, all, and *kreas*, flesh.

97. The Stomach and Intestines are each composed of three coats. The outer (*serous*) coat is smooth and glistening; the middle (*muscular*) coat is really a double membrane, one set of fibres are longitudinal, the other circular; the interior (*mucous*) coat is covered with *Villi* (hair-like projections), which give it a velvety appearance. From these projections the *Lacteals* (or vessels that receive the digested fluids) arise. (See p. 133.) Many parts of this membrane are doubled upon itself, forming folds. (Fig. 51.)

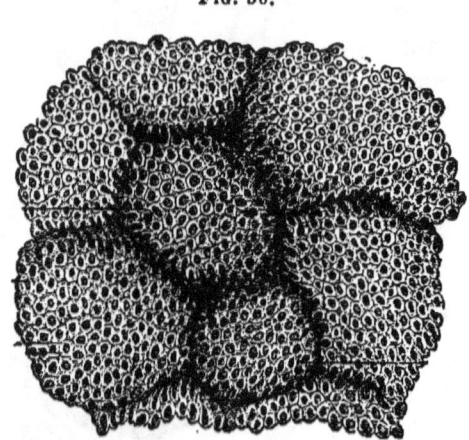

FIG. 50.

FIG. 50 EXHIBITS THE ORIFICES of the Gastric Glands of the Mucous Membrane of the Stomach.

98. The Alimentary (food) Canal is lined its entire length by *mucous membrane*, which is composed of three layers, the surface layer (*epithelium*), the *basement* membrane and the cellular (*connective*) layer. (Fig. 3.) In different parts of the alimentary canal the surface layer varies in the form of its cells. In the mouth they are flattened, resembling thin scales; in the upper part of the throat the cells at their free extremity are fringed or hair-like. During life these fringed processes are endowed with a power of moving rapidly backward and forward in a wave-like manner, reminding one of the movement of a field of grain swept by a gentle breeze. Currents are thus produced in liquids, conveying them from one part to another. In the stomach and intestines the cells are six-sided columns.

How many coats have the Stomach and Intestines? Describe the serous coat. The muscular coat. The interior coat. By what is the alimentary canal lined? Of what is this membrane composed? State the appearance of the epithelial cells in different parts of the alimentary canal.

THE DIGESTIVE ORGANS. 57

Observation 1.—Epithelial cells not only exist in the mucous membrane of the alimentary canal, but in the bronchial membrane of the lungs, in serous and synovial membranes, in the skin, the ducts of all glands, the ventricles of the brain, the canal of the spinal cord, the absorbents and the inner coat of the blood-vessels. From the extent and connection of these cells we learn the peculiar transfer of diseases from one part of the body to another.

2.—Diphtheria is a diseased condition of the epithelial cells of the mouth and throat. Irritation of the epithelium of the stomach induces vomiting; a morbid state of these cells of the small intestine causes diarrhœa; an inflammatory action of the large intestine (the rectum), dysentery. In these diseases it is always safe to invite the blood to the skin by bathing, friction and extra coverings, to induce free and continued perspiration.

§ **11.** Physiology of the Digestive Organs.—*The Assimilation of Food. Process by which Food is changed into Chyle. Destination of the Chyle.*

99. Food is necessary to the preservation and growth of the body, but it must first be changed into matter having the *same characteristics* as those organs and tissues that it is destined to nourish.

The change of both the solids and fluids taken into the body as food into a milky, nourishing fluid may be termed *Primary* (first) *Assimilation*. The change of the blood necessary to meet the repair and waste of the body may be termed the *Secondary Assimilation*. (See p. 130.)

Fig. 51.

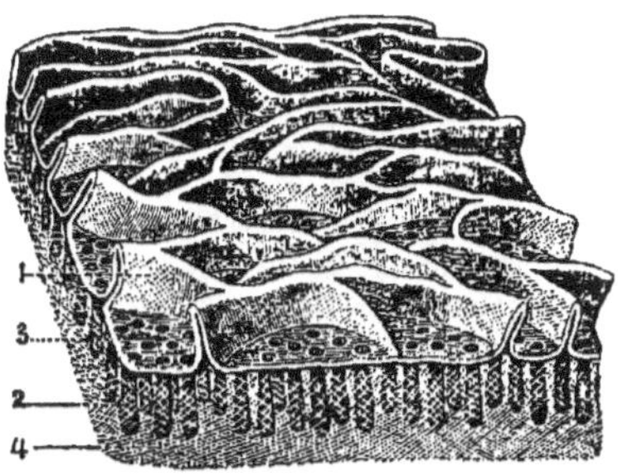

Fig. 51. Mucous Membrane from the Jejunum.—1, Villi (folds of lining mucous membrane) in miniature. 2, Tubular glands: their orifices. 3, Opening on the free surface of the mucous membrane. 4, Fibrous tissue. Magnified.

100. The chemical or hidden processes concerned in digestion consist of peculiar changes (*reactions*) between the food and the various secretions of the alimentary canal. These

---

Observation. What change in food is necessary? What is Primary Assimilation? What Secondary?

fluids are—*mucus* and *saliva*, secretions of the mucous membrane and glands of the mouth; *gastric juice*, a secretion of the stomach; *bile*, a secretion of the liver; *pancreatic juice*, a secretion of the pancreas; *mucus* and *intestinal juices*, secretions of the mucous membrane and glands of the intestines. Each of these fluids effects a special change in the food.

101. The alimentary canal in which these digestive changes take place is like a long manufacturing establishment with many apartments; the first room being the Mouth, or masticating room, where some of the workmen cut the food; some grind it; some moisten it and supply the needed chemicals for making one of the changes. Mastication being completed,

Fig. 52.

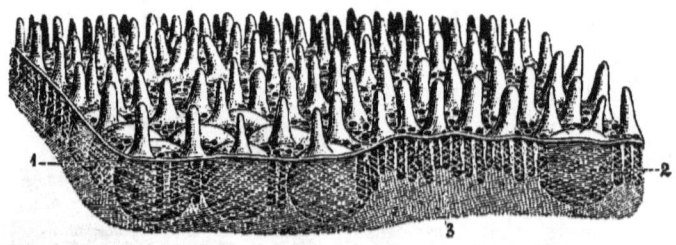

Fig. 52. Portion of the Mucous Membrane from the Small Intestine, magnified, showing the villi on its free surface, and between them the orifices of the tubular glands.—1, Portion of a clustered gland. 2, A solitary gland. 3, Fibrous tissue.

at the word of command the obedient muscles, with greatest promptness and efficiency, convey the food onward to that wonderful laboratory, the Stomach.

102. In this second apartment, the Stomach, the food, by the action of the muscular coat of this organ, is subjected to a churning or rotary motion which brings part after part of the food in contact with a remarkable chemical agent, the *Gastric Juice*, which changes it into a soft pulp called *chyme*.

Observation.—While digestion is thus going on, the openings of the stomach are well guarded. A return of any part of the mass into the œsophagus is prevented by the sphincter muscles near the cardiac orifice, and the passage to the intestine is closed by the sphincter mus-

---

Name the fluids that effect a chemical change in food. To what is the alimentary canal likened? Speak of the process of the digestion of food in the Stomach. Observation.

cles of the pyloric orifice and a valve called the pylorus or "gate-keeper," which, true to its name, stands a faithful sentinel till proper chyme presents itself.

103. The CHYME when perfected passes into the third room, the upper portion of the small intestine (Duodenum). Here this greyish, pulpy mass is treated by other chemical agents, the bile, the pancreatic and intestinal juices, by which it is converted into a whitish, semi-fluid substance called *Chyle*. By a wave-like (*peristaltic*) muscular action of the intestines the pulp is forced to its respective destination. The nutritive portion is taken up by the *Lacteal Absorbent* vessels, and the waste portion (the innutritious) is passed into the large intestine and excreted from the body.

Observation.—The absorbing surface of the intestines is greatly increased by the projecting forms and great abundance of the *villi;* they hang out into the nutritious, semi-fluid mass contained in the cavity of the intestines as the roots of a tree penetrate the soil, and imbibe the liquid portions of food with wonderful rapidity.

§ **12.** HYGIENE OF THE DIGESTIVE ORGANS.—*Suggestions relative to the Preservation of the Teeth—To their Removal. Conditions affecting the Quantity of Food demanded by the System—The Quality of Food. Directions relating to the Manner of taking Food. Conditions of the System requisite for the proper Digestion of Food.*

104. *For the preservation of the Teeth* the first requisite is to keep them clean. After meals they should be cleansed to prevent the collection of *tartar* and to remove any remaining particles of food.

105. *The removal of the Teeth.* The temporary teeth should be removed at once when loose, and often before, when the permanent teeth appear. This is essential to the regularity and beauty of the second set.

Observation 1.—Irregular or crowded permanent teeth generally require the removal of one or more. By pressure upon each other the enamel is injured and the appearance rendered unsightly.

What changes in the food occur in the Small Intestine? What becomes of the nutritive portion of the food? The innutritious? How is the absorbing surface of the intestine increased? Name the first requisite for the preservation of the Teeth. When should the temporary teeth be removed? What do the irregular permanent teeth generally require?

2.—Toothache does not always indicate the necessity of extraction, as the nerve or investing membrane may be diseased and the tooth sound. When the removal of a tooth is necessary, apply to some skillful operator; something more is needed than strong muscles and a pair of forceps. Skill is as requisite in the proper extraction of a tooth as in the amputation of a limb.

106. The health of the Digestive Organs in general requires the observance of certain conditions relative to their natural stimulus, *Food.* These will be considered under the following heads: 1. *The Quantity of Food.* 2. *The Quality of Food.* 3. *The Manner of taking Food.* 4. *The Proper Conditions of the System for receiving Food.*

107. The QUANTITY OF FOOD necessary to the system varies, being affected by age, occupation, temperament, habits, temperature, amount of clothing, health and mental state.

108. *The supply must equal the waste of the system.* In every department of nature waste attends action. The greater the amount of exercise the more rapidly will the particles be worn out and removed and their places need supplying with new atoms.

Observation 1.—During the period of growth the supply must exceed the waste for the building of new tissues. This accounts for the keen appetite and vigorous digestion in childhood. The same is true when persons have become emaciated from famine or disease.

2.—When exercise is lessened, the quantity of food should be proportionally diminished, otherwise the tone of the digestive organs will be impaired and the health of the system enfeebled. This is especially applicable to students who have been accustomed to laborious employments.

109. *More food is required in winter than in summer;* hence by diminishing the amount of food as the warm season approaches, the tone of the stomach and vigor of the body will be better maintained, thus lessening the liability to "summer complaint."

Observation.—The amount of food should be adapted to the present condition of the digestive organs. Imperfectly digested food irritates

---

Observation 2. What is required for the health of the Digestive Organs? What is said of the quantity of food? What must the supply equal? When must supply exceed waste? When should the quantity of food be diminished? Is more or less food required in winter than in summer? To what should the amount be adapted?

the mucous membrane of the intestines and enfeebles the system instead of invigorating it. In sickness the attending physician is the person to decide respecting the proper amount, as diseased tissues suffer from undue action.

110. The QUALITY OF FOOD *should be both nutritive and digestible.* Substances are *nutritious* in proportion to their capacity to yield the elements of chyle. Substances are *digestible* in proportion to the facility with which they are acted upon by the digestive fluids. Articles highly nutritive in themselves, but difficult of digestion, often yield less nourishment than those poorer in nutritive quality but easy of digestion.

Observation.—If we confine our diet to easily digested articles, the digestive organs will be weakened from want of proper exercise; if too highly concentrated diet, they will be injured by over-work; hence the necessity of choosing, in this respect, the "happy medium." Variety in food is as essential in the domestic animals as in man.

111. *Food should be properly cooked.* However nutritious an article of food may be, if not well cooked it is not only unsavory to the palate but hurtful to the digestive organs.

Observation.—The simplest methods of preparation by cooking are the best. Meat should be broiled, roasted or made into soup. The cooking of *vegetables* should be thorough and complete.

112. *The Quality of Food should be adapted to the season and climate.* Highly stimulating food may be used almost with impunity during the cold season of a cold climate, but in the warm season and in a warm climate it is very injurious. Animal food, being more stimulating than vegetable, is therefore well adapted to winter, and vegetable to spring and summer. Where the digestive organs are weakened or diseased, it is very important that a nutritious vegetable diet be adopted as the warm season approaches.

Observation.—Vegetable diet is most suitable for children. The organs of a child are more sensitive and excitable than those of an adult; hence, stimulants of every kind should be strictly avoided, and the food

---

What should be the quality of food? Distinguish between nutritious and digestible substances. Observation. How should food be cooked? To what should the quality be adapted? What is said of vegetable diet?

mainly of a vegetable character. In this "fast age" this is a suggestion of vast importance.

113. The MANNER OF TAKING FOOD exercises a controlling influence upon the health of the digestive organs. It is essential that the *food be properly masticated* to secure the fine division necessary to the proper action of the gastric juice and other fluids, and especially to mix the food with the requisite amount of saliva.

Observation.—Rapid eating should be avoided, not only as a violation of good table manners, but as a violation of the laws of our physical nature, whose penalty, in the form of dyspepsia with its numerous train of evils, will sooner or later be visited upon the transgressor.

Drink should not be taken in excess with the food. Nature supplies the appropriate moisture, and if tea, coffee or any other fluid be used as a substitute, indigestion may follow from the absence of the necessary amount of saliva. Again, drinks taken into the stomach must be absorbed before the digestion of other articles is commenced.

114. *The temperature of food and drink should be observed.* Hot food or drink for a short time unduly stimulates the vessels of the mucous membrane of the gums, mouth and stomach; then reaction follows, bringing loss of tone and debility of these parts. This practice is a fruitful cause of spongy gums, decayed teeth, sore mouth and indigestion. On the other hand, if food or drink be taken too cold, or iced, an undue amount of heat is abstracted from the stomach; this arrests the digestive process, and thus deranges the system.

Observation.—The temperature of the stomach during digestion is from 98° to 101°. When *iced* food or drinks are taken, digestion does not commence or continue until they are warmed to 98° or removed from the stomach.

115. *The food should be taken at regular and suitable periods.* The interval between the meals should be regulated by the character of the food, and the age, health, exercise and habits

---

What is said of the manner of taking food? Why should food be properly masticated? Why not take drink in excess with food? Why should regard be had to the temperature of food and drink? Observation. How and when should food be taken?

of the individual. In the young, the active and the vigorous, food is more rapidly digested than in the aged, the indolent and the feeble; consequently, it should be taken more frequently by the former class than by the latter.

116. The CONDITIONS OF THE SYSTEM FOR RECEIVING FOOD are important in digestion.

Food should not be taken immediately before or after severe exercise of body or mind. The functional exercise of any organ takes the fluids from other parts of the body, thus weakening those parts for the time. Severe exercise of the muscle concentrates the forces in the muscle; severe exercise of the brain concentrates the forces of the brain; the same is true of the vocal and other organs. After severe exercise, from thirty to forty minutes should be allowed before eating, for restoring equilibrium to the system.

Observation 1.—The student, farmer or mechanic who hurries from his toil to his dinner to "save time" will in the end lose more time than he saves. After eating, the digestive organs need, for a time, the chief use of the vital forces, and if they are habitually expended elsewhere, as in study or labor, digestion will be arrested, the chyle cheated of its proper elements, and headache, dullness and general derangement will follow.

2.—When horses and oxen have been worked hard, water or food should not be given as soon as they are stabled. The noon meal of the worked domestic animal should be light. Neither *water* nor food should be given the hard-driven horse until he is somewhat rested.

117. *Persons should abstain from eating at least three hours before retiring for sleep.* It is no unusual occurrence for those persons who have eaten heartily immediately before retiring to have unpleasant dreams, or to be aroused from their unquiet slumber by colic pains. In such instances the brain becomes partially dormant, not imparting to the digestive organs the requisite amount of nervous influence; this being deficient, the unchanged food remains in the stomach, causing irritation of this organ.

Observation.—A healthy farmer who was in the habit of eating a

---

State the reason for not taking food just before or after exercise. Observation. **Why** is it not best to eat immediately before retiring to sleep? Observation.

quarter of a mince pie just before retiring became annoyed with unpleasant dreams, and among the images of his fancy he saw that of his deceased father. Becoming alarmed, he consulted a physician, who advised the patient to eat *half* a mince pie, assuring him that then he would *see his grandfather*.

118. *The state of the mind exerts an influence upon the digestive process.* This is clearly shown when an individual receives sad intelligence. Let him be sitting at a plentiful board with a keen appetite, and the unexpected news destroys it, because the excited brain withholds the stimulus; hence, all unpleasant themes, labored discussions or matters of business should be banished from the table. Light conversation, enlivening wit and cheerful humor wonderfully promote digestion.

Observation 1.—Indigestion arising from nervous prostration should be treated with great care. The food should be simple, nutritious, properly cooked, moderate in quantity and taken at regular periods. Large quantities of stimulating or rich food, frequently taken, serve to increase the nervous prostration. Exercise in the open air and a cheerful state of mind are very beneficial in restoring the natural, healthy action of the brain, and thus aiding the digestive powers.

2.—After long abstinence light, nourishing food should be taken, and in small quantities. As in case of sickness, when the appetite begins to return, the nurse must use much discretion, and the patient, often, self-denial. The popular adage that "food never does harm where there is a desire for it" is untrue. Too frequently, when a patient satisfies his cravings, it is to induce a relapse into the former disease, and at the risk of life. The digestive organs are weak, and must be gradually brought into action. It is often better to give the food in a solid rather than liquid form, so that the salivary and mucous glands may be stimulated to action.

119. *The conditon of the skin exercises an important influence upon digestion.* Let free perspiration be checked, either from uncleanliness, chills or any other cause, and the functional action of the stomach is diminished. This is one of the fruitful causes of "liver and stomach complaints" among the filthy and half-clad inhabitants of our cities and villages.

What influence does the state of the mind exert upon the digestive organs? How should indigestion arising from nervous prostration be treated? After long abstinence, what kind of food should be taken? What influence does the condition of the skin exert?

Attention to bathing and clothing would prevent many "season complaints," especially among children.

Observation.—The useful cow should be protected from chilling rains and frosts. It is poor economy to have the skin of any domestic animal chilled.

120. *Pure air is necessary to give a keen appetite and vigorous digestion.* The digestive organs must have a plentiful supply of pure blood, and to have pure blood we should breathe pure air. Poor ventilation is a frequent cause of indigestion. Persons who sleep in ill-ventilated rooms have little or no appetite in the morning.

General Observations.—A manufacturer stated before a committee of the British Parliament that he had removed an arrangement for ventilating his mill, as he noticed that his men ate much more after his mill was ventilated than before, and he could not *afford* to have them breathe the pure air. Compression of the vital organs prevents the introduction of a sufficient supply of pure air, and is one of the causes of dyspepsia, now so prevalent among ladies.

All aliment is separated into nutriment and residuum. The latter should be regularly expelled from the system; otherwise headache, dizziness and general uneasiness will ensue, and if allowed to continue the foundation will be laid for a long period of suffering and disease. For the preservation of health there should be in most persons a daily evacuation of residual matter. Evening is the best time; especially is this true when persons are afflicted with piles. Constipation may in many cases be relieved by friction over the abdominal organs and by making an effort to evacuate the residuum at some stated period each day.

Recapitulation.—Digestion is most perfect when the action of the skin is energetic; the brain moderately stimulated; the blood well purified; the muscular system duly exercised; the food properly cooked and masticated, taken at regular periods, and adapted in quality and quantity to the present condition of the individual.

§ 13. COMPARATIVE ANATOMY (Splanchnology).—*Nutritive Apparatus of Vertebrates. Compare the Mouths and Teeth of Vertebrates— The Stomach and Intestines of Vertebrates—Nutritive Apparatus of Annulosa and Mollusca—Of Radiata—Of Protozoa.*

121. In the NUTRITIVE APPARATUS of all Vertebrates a

Observation. Why is pure air necessary? General Observation. Recapitulation.

general plan of parts obtains, subject to the variations required to preserve the harmony of relation between the organization and the use to which it is to be applied.

122. In no part do we find a greater variety or a nicer accommodation to particular wants than in the MOUTHS and TEETH of different animals. In *Mammals* the projecting jaws, the wide mouth, the strong, pointed, sharp, enameled edges of the teeth enable flesh-eating animals to seize and hold their prey, and the hinge-like movement of the jaw to divide it like a pair of scissors, as seen in the Cat and Lion. (Fig. 53.) The full lips, the rough tongue, the furrowed,

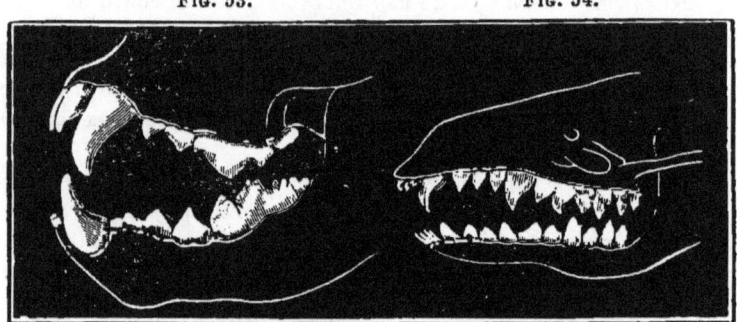

FIG. 53. FIG. 54.

FIG. 53. THE TEETH OF CARNIVORA OR FLESH-EATING ANIMALS.
FIG. 54. THE TEETH OF INSECTIVORA OR INSECT-EATING ANIMALS.

cartilaginous palate, the broad, rough surface of the teeth, the central plates of enamel and the lateral movement of the jaw qualify the herbivorous or grain-eating animals for grazing and for grinding their food, as grain is crushed between the upper and nether millstone, as in the Sheep and Horse.

The elongated, tapering muzzle, the cone-pointed, enameled molars locking into the enameled depressions of the opposite jaw, enable the insectivorous animals to burrow in the earth for the insects and worms upon which they feed, and also to crush them, as in the Mole and Hedge-hog. (Fig. 54.)

What is said of the Nutritive Apparatus of Vertebrates? Speak of the mouth and teeth of Mammals—as the Lion and Cat,—the Horse and Sheep,—the Mole and Hedgehog,—the Rat and Squirrel.

The two chisel-shaped incisors, enameled only in front, allowing more rapid wear of the posterior than the anterior part, keeping them always sharp; the bag of pulp at the base of these teeth, providing for growth equal to the wear at the top; the backward and forward movement of the jaws and the great size and strength of the lower jaw, adapt the rodentia or gnawers to their mode of life, as in the Rat and the Squirrel. (Fig. 55.)

In *Birds* the mouth receives a new character, both in substance and in form. Instead of fleshy lips and teeth of enameled bone, we have the hard and horny investment of the jaws, known as the *bill,* destitute of true teeth. This organ

FIG. 55.

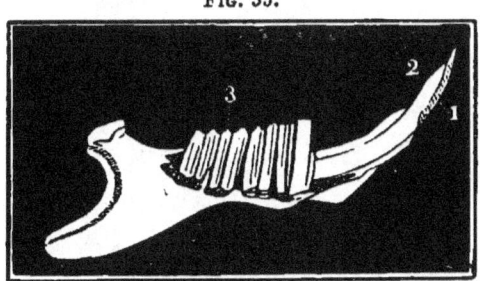

FIG. 55. LOWER JAW OF A SQUIRREL.—1, The enamel of the gnawing tooth. 2, The ivory. 3, The lateral furrows of the molar teeth.

varies in size and form according to the food of the species, which may be grains, insects, fishes or flesh.

In all carnivorous *Reptiles* the prey is swallowed whole; hence their jaws and throats are made capable of great dilatation. Their teeth are used only for seizing and retaining their prey, but not in any way for dividing it.

Some species of *Amphibians,* as Frogs, have only the upper jaw armed with teeth. The structure of the tongue of the Toad is like that of the Frog (attached to the floor of the mouth), but the jaws are not furnished with teeth.

The teeth of *Fishes* vary much in form in different species, being sometimes fine and thickly set; in others they are strong hooks or sharp-cutting plates.

123. The STOMACH and INTESTINES of *Mammals* vary in size, form and relative length. They are simpler, harder and shorter in flesh-eating than in herbivorous or grain-eating animals; while the Ox has intestines about twenty times the length of his body, those of the Lion are but three or four times its own length.

Ruminants (those animals that chew the cud), as the Sheep and Ox, have a stomach with four cavities. The first stomach, called the *Ru'men* (Paunch); the second, the *Retic'u-*

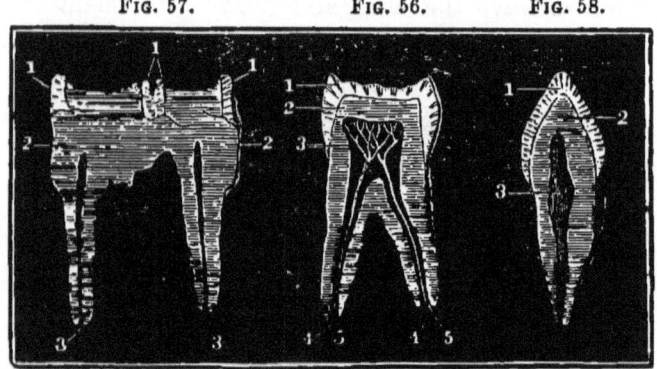

FIG. 56. A SECTION OF A MOLAR TOOTH OF MAN.—1, The enamel. 2, The ivory. 3, The cavity containing blood-vessels. 4, 5, Artery and nerve.

FIG. 57. SECTION OF THE MOLAR TOOTH OF A HORSE.—1, The enamel. 2, The ivory. 3, Canal for blood-vessels.

FIG. 58. SECTION OF MOLAR TOOTH OF A DOG.—1, The enamel. 2, The ivory. 3, Cavity for blood-vessels. Diagrams.

lum (Honeycomb); the third, the *Oma'sum* (Many-plies); the fourth, the *Ab'omasum* (Rennet); the latter, taken from the young calf, is used in cheese-making. (Figs. 59, 60.)

The food when first swallowed is received into the Rumen, where it accumulates while the animal is feeding. Here it is moistened by the fluids secreted by the walls of this cavity. It then passes into the Reticulum, where it receives additional secretions and is made into little pellets or "cuds,"

Speak of the stomach and intestines of Mammals. Vertebrates. Describe the stomach and give the process of digestion in Ruminants.

THE DIGESTIVE ORGANS.

which, when the animal is at rest, are returned to the mouth to be re-chewed and mixed with the saliva. This pulp passes directly into the third cavity to be prepared for the fourth, where digestion is finally completed. It is then received by the intestinal canal.

In *Birds* there are usually three cavities or stomachs; the first is an expansion of the œsophagus, called the *Crop* (*Inglu'vies*), where the food is macerated and softened; the second is the true stomach (*Proventric'ulus*), where the mucous mem-

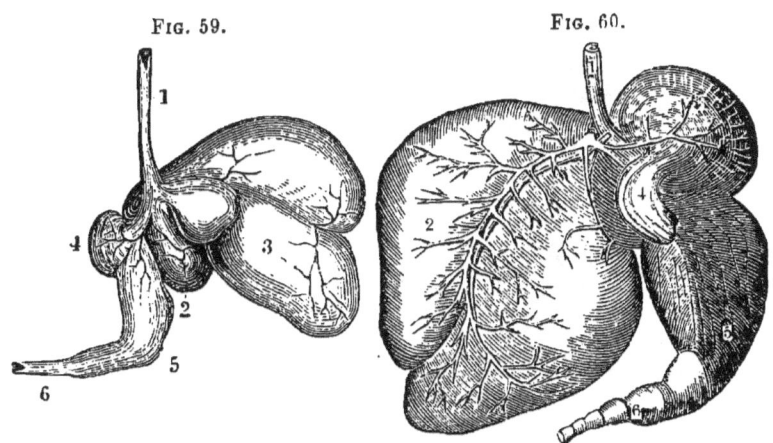

FIG. 59. STOMACH OF THE SHEEP.—1, The œsophagus. 2, The rumen. 3, The reticulum. 4, The omasum. 5, The abomasum or rennet. 6, The intestine.

FIG. 60. STOMACH OF AN OX.—1, The œsophagus. 2, The rumen (paunch). 3, The reticulum (honeycomb). 4, The omasum (many-plies). 5, The abomasum (rennet). 6, The intestine.

brane is provided with mucous cavities, secreting an acid which acts still farther upon the food; and the third is the *Gizzard* (*Trit'urating cavity*), which leads into the commencement of the small intestine. The latter, in grain-eating birds, has immense strength, being composed of muscular fibres running in different directions and lined with a horny membrane. Gravel and angular stones are instinctively swallowed to assist

---

Name and describe the stomachs of Birds.

70 ANATOMY, PHYSIOLOGY AND HYGIENE.

in the grinding process. In flesh-eating birds the gizzard is thin and membranous. The commencement of the large in-

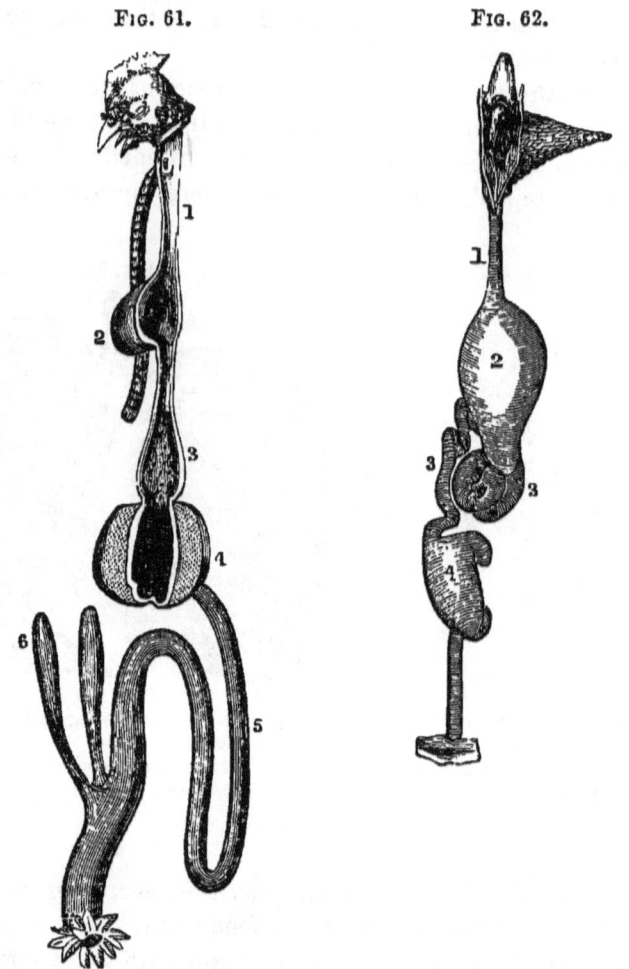

FIG. 61. THE ALIMENTARY CANAL OF A FOWL.—1, The œsophagus. 2, Ingluvies (crop). 3, Proventriculus (secreting stomach). 4, Triturating stomach (gizzard). 5, Intestine. 6, Two cæca.

FIG. 62. THE ALIMENTARY CANAL OF THE FLYING LIZARD.—1, The œsophagus. 2, The stomach. 3, 3, Small intestine. 4, Large intestine.

testine is furnished in most birds with two blind tubes or cæca. Their exact function is still questionable. (Fig. 61.)

THE DIGESTIVE ORGANS. 71

Observation.—It is essential to the health of domestic fowls, when kept in enclosures, to supply them with gravel or small pebbles to act as teeth in grinding their food. It is poor economy not to give hens light, pure air, a supply of pure water and feed regularly, with warm apartments.

In *Reptiles* the alimentary canal differs much from that of mammals or birds. As a general rule, it is shorter in proportion to the trunk than in warm-blooded vertebrates. The passage from the œsophagus to the stomach is by a pouch-like enlargement. The small intestines usually have a few coils; the large intestines in most reptiles are short, simple and straight, without cæcal appendage at its beginning. The liver is relatively large. (Fig. 62.)

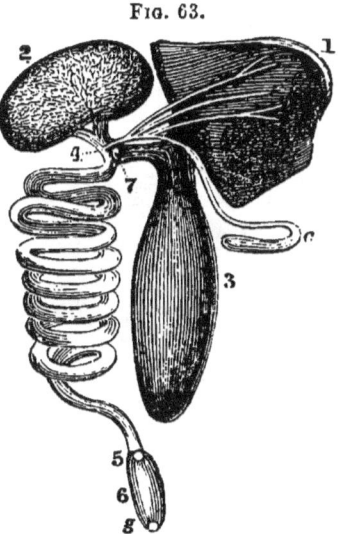

FIG. 63.

FIG. 63. THE ALIMENTARY CANAL OF THE SWORD-FISH.—1, Liver. 2, 3, Cæcas or pouches connecting with small intestines. 4, 5, Small intestine, coiled. 6, Large intestine. 7, Biliary duct.

In *Fishes* the alimentary canal is more diversified in length, size and form than in reptiles. There are two predominant forms of the stomach in fishes—one like a bent tube (siphonal), and the other a blind tube (cæcal). In some species of fish the small intestines extend in a line from the stomach to their termination; in others there are found from two to eight coils. The large intestines are short and straight, and the termination of the rectum opens into a cavity called the *Cloaca*. The liver is usually large, with numerous appendages. In the cod it is soft and saturated with oil, which is expressed for medicinal purposes. (Figs. 63, 64.)

124. The object of digestion in Invertebrates as well as Vertebrates is to separate the nutritious part of the aliment

---

Observation. Speak of the digestive organs of Reptiles. What is said of the stomach and alimentary canal in Fishes? What is the object of digestion in Invertebrates?

from the innutritious portion of the residuum, so that the former may be converted into liquids adapted to mingle with the blood.

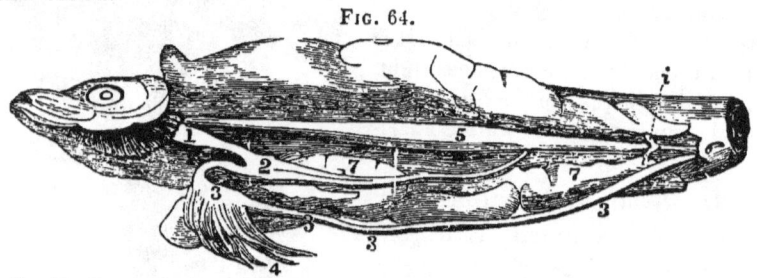

FIG. 64. THE ALIMENTARY CANAL OF THE HERRING.—1, The œsophagus. 2, Stomach. 3, 3, 3, 3, Small intestine. 4, Cæca. 5, Air-bladder. 7, Pneumatic duct.

125. The ANNULOSA and MOLLUSCA are furnished with a distinct alimentary canal that does not open into the body-cavity. In most cases the digestive canal communicates with the outer world by two openings—a mouth and an excretory aperture. (Fig. 65.)

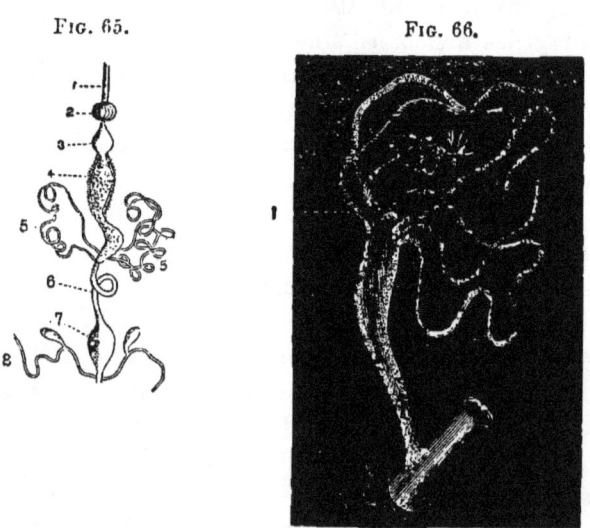

FIG. 65. DIGESTIVE APPARATUS OF A BEETLE.—1, Gullet. 2, Crop. 3, Gizzard. 4, Chylific stomach. 5, Malphigian tubes or cæca. 6, Intestines with cloaca. 8, Renal vessel.
FIG. 66. THE DIGESTIVE APPARATUS of the Hydra or Fresh-water Polyp.

Speak of the digestive organs in the Annulosa and Mollusca.

THE DIGESTIVE ORGANS.

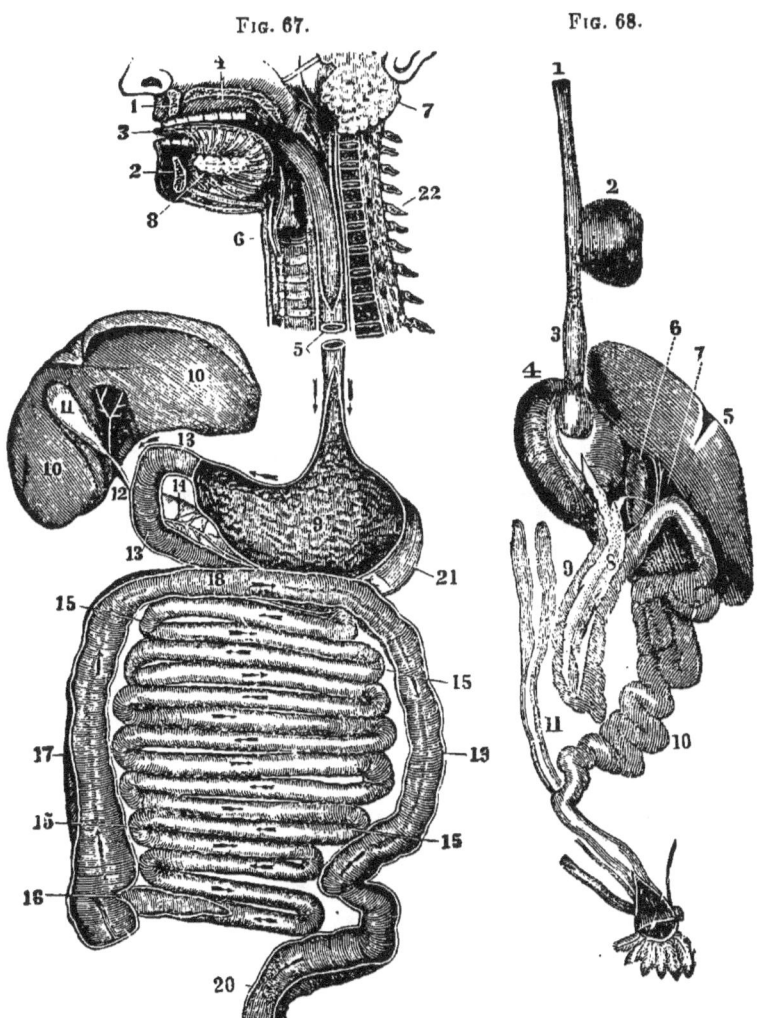

FIG. 67. DIAGRAM OF THE ORGANS OF DIGESTION, opened nearly the whole length.—1, The upper jaw. 2, The lower jaw. 3, The tongue. 4, The roof of the mouth. 5, The œsophagus. 6, The trachea. 7, The parotid gland. 8, The sublingual gland. 9, The stomach. 10, 10, The liver. 11, The gall cyst. 12, The duct that conveys the bile to the duodenum (13, 13). 14, The pancreas. 15, 15, 15, 15, The small intestine. 16, The opening of the small intestine into the large intestine. 17, 18, 19, 20, The large intestine. 21, The spleen. 22, The upper part of the spinal column.

74 ANATOMY, PHYSIOLOGY AND HYGIENE.

126. In the RADIATA the digestive cavity is a pouch with a single opening, into which the food is passed and from which the residuum is ejected, as in the Hydra. (1–Fig. 66.)

127. In the PROTOZOA there is no digestive apparatus, or only a rudimentary one. The process of nutrition is carried on in the simplest possible manner and with the simplest possible apparatus. The only distinct structure which is at all concerned in nutrition is a contractile cavity which opens and closes at definite intervals.

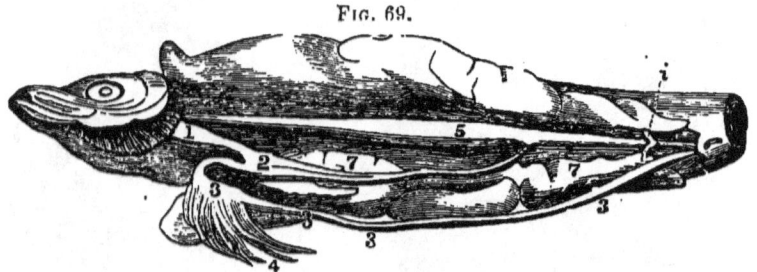

FIG. 69. THE ALIMENTARY CANAL OF THE HERRING.—1, The œsophagus. 2, Stomach. 3, 3, 3, 3, Small intestine. 4, Cæca. 5, Air-bladder. 7, Pneumatic duct.

Speak of the digestive organs in the Radiata. In the Protozoa.
State the Anatomy, the Physiology, the Hygiene, Human and Comparative, of the Digestive Organs.

SYNTHETIC TOPICAL REVIEW.
DIGESTIVE ORGANS.

Classes, Sub-kingdoms, Divisions, Anatomy, Physiology and Hygiene.

CHAPTER V.

ABSORPTION.

128. THE changing of food into chyle in the Digestive Organs is virtually *external* to the animal body. The process by which it is conveyed *within* to enrich the blood, and also the taking up or removal of the parts of living tissues within the body, is called *Absorption,* and the vessels conveying it are named *Absorbents.*

§ **14.** ANATOMY OF THE ABSORBENTS.—*The Absorbent Vessels. Distribution of the Lymphatics. The Thoracic Duct. The Lymphatic Duct. Lacteals. Lymphatic Glands. Absorbent Veins.*

129. The ABSORBENTS consist of certain blood-vessels, especially the venous capillaries and the absorbents proper, viz., *Lymphatic* * *Vessels* and *Glands.* The lymphatic vessels of the small intestines are named *Lac'teals.* †

130. Most of the LYMPHATIC VESSELS are long, thread-like, transparent tubes, with coats exceedingly delicate. They are distributed through most of the system. Few are found in the muscles and none in the brain or spinal cord, though they doubtless exist there. They abound in the secreting membranes, especially in the skin and the mucous membrane.

The finer lymphatics unite into trunks, which either accompany the blood-vessels and form the *deep* lymphatics, or run on the surface of organs, forming *superficial* lymphatics. From all parts of the body these trunks run toward the root of the neck and unite in two main trunks, called the *Thoracic* and *Lymphatic Ducts.* (Fig. 77.)

131. The THORACIC DUCT is formed by the uniting of the lymphatic vessels from the lower extremities, those of the *left*

What is Absorption? Define Absorbents. Of what do the Absorbents consist? Describe the Lymphatic Vessels. Speak of the Thoracic Duct.

* Lat., *lympha*, water. † Lat., *lac*, milk.

76 ANATOMY, PHYSIOLOGY AND HYGIENE.

Fig. 72.

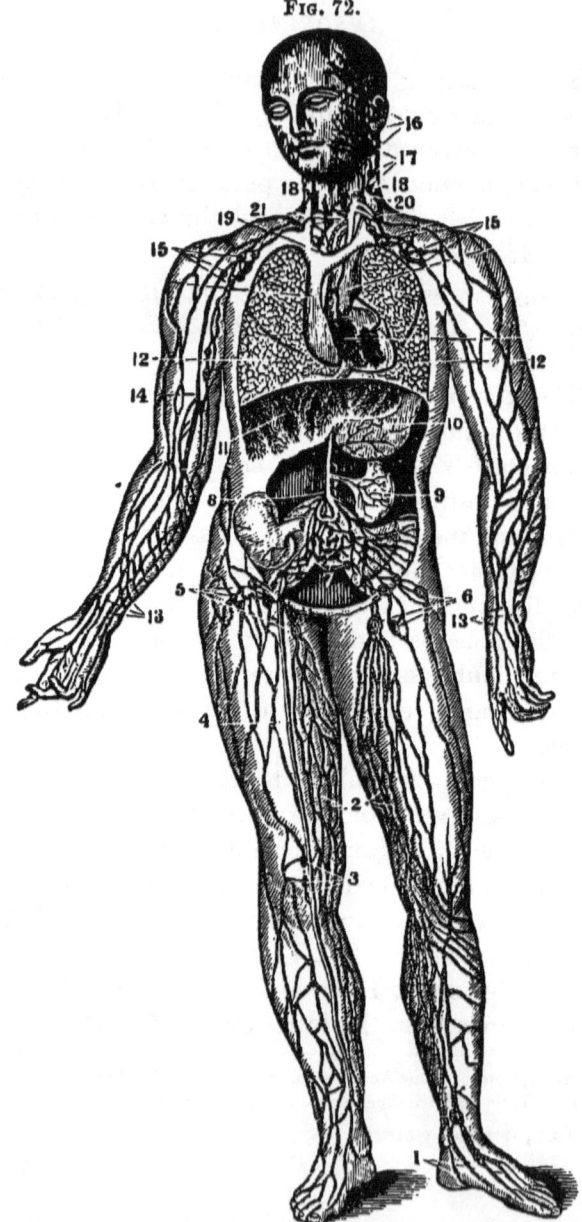

FIG 72. THE LYMPHATIC VESSELS AND GLANDS.—1, 2, 3, 4, 5, 6, Lymphatic vessels and glands of the lower limbs. 7, Lymphatic glands. 8, The receptacle of the thoracic duct. 9, The lymphatics of the kidney. 10, Of the stomach. 11, Of the liver. 12, 12, Of the lungs 13, 14, 15, The lymphatics and glands of the arm. 16, 17, 18, Of the face and neck. 19, 20, Large veins. 21, The thoracic duct.

side of the head and neck and left upper limb, also those of the abdomen. It commences behind the liver and ascends in front of the spinal column. At the lower part of the neck it turns downward and forward, and pours its contents into the vein behind the left collar-bone. The duct is equal in diameter to a goose-quill. (Figs. 72, 76.)

132. The LYMPHATIC DUCT is about an inch long. It is formed by the union of the lymphatic vessels of the *right* side of the head and upper extremities, and terminates in a vein in the right side of the neck. (Fig. 72.)

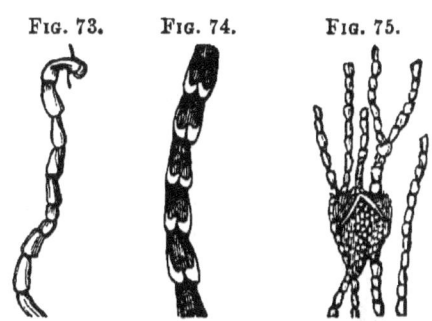

FIG. 73. A SINGLE LYMPHATIC VESSEL, much magnified.

FIG. 74. THE VALVES of a lymphatic trunk.

FIG. 75. A LYMPHATIC GLAND, with several vessels passing through it.

133. The LACTEALS commence in the internal coat of the small intestine. These minute vessels unite and reunite with each other and pass through small glands (*Mes'enteric**) to the Thoracic Duct.

134. The LYMPHATIC GLANDS are not well understood. They seem to be composed of a large number of hard, pinkish bodies, varying in size from that of a hemp seed to that of a large pea, which communicate with each other and also with the lymphatic tubes. (Figs. 72, 75.)

Observation.—The lymphatic glands are found in the axilla of the arm (arm-pit) and in the groin; chains of glands are found on each side of the neck; a few in the arm; also many about the bronchi or air-tubes and in the pelvis or abdomen, those of the lacteals being abundant in the *Mes'entery*.

135. The VEINS of the intestines acting as absorbents unite with those coming from the stomach, the spleen and the pan-

---

The Lymphatic Duct. The Lacteals. Describe the Lymphatic Glands. Where found?

* Gr., *mesos*, middle, and *enteron*, the intestine.

creas, thus forming the *Portal vein,* which enters the liver on the under surface. (Fig. 112.)

15. PHYSIOLOGY OF THE ABSORBENTS.—*Lymph. Veins as Absorbents. Absorbent Power of different Membranes. The Effect of Inactivity of the Absorbents. Absorption by the Skin and other Membranes in cases of Disease. Absorption in the Inferior Animals.*

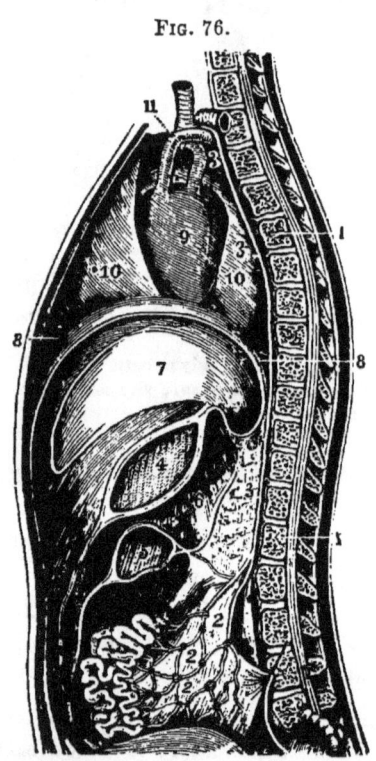

FIG. 76. SIDE VIEW OF THE LACTEALS AND THORACIC DUCT.—1, Small intestine. 2, 2, 2, Lacteals. 3, 3, 3, Thoracic duct. 4, Stomach. 5, Colon. 6, Pancreas. 7, Liver. 8, 8, Diaphragm. 9, Heart. 10, 10, Lungs. 11, Large vein into which the thoracic duct opens. 12, 12, Spinal column.

136. We know little of the changes which take place in the Absorbents. Chyle drawn from the Thoracic Duct is very different from that just absorbed by the lacteals.

137. The fluid (*Lymph*) which circulates through the lymphatics of the limbs is clear and colorless, and differs from the milky chyle. Lymph, like chyle, is now considered a nutritive fluid.

Observation.—There is much evidence that the lymph is obtained from the blood, and it is not improbable that the lymphatics take up those crude materials which were absorbed directly by the veins and subject them to an assimilating agency resembling that acting upon the nutritive substances in the lacteals. The whole lymphatic system may be looked upon as one great assimilating or blood-making gland.

138. Many if not all the *Veins* are Absorbents. As every respiration, every heart-beat, every muscular movement, every thought, is produced

---

What is the Portal vein? What is known of the function of the Absorbents? Speak of Lymph. Observation.

at the expense of the life of some of the tissues, the special office of the veins as absorbents seems to be to take up or remove the waste particles no longer of use to the living tissues and convey them from the body.

139. Different membranes have different absorbent powers, and the power of the same membrane varies with change of condition. The most active is the mucous membrane; thus, in the alimentary canal it takes up a large portion of the food; in the lungs it absorbs gases in a state of solution. In this way are introduced into the system miasmatic and contagious exhalations. Fine, solid particles are sometimes absorbed, as arsenic.

140. There are no visible openings in the membranes for the passage of these absorbable substances, but their entrance seems to be effected by a peculiar action of animal membranes, which enables certain fluids to pass directly through them.

Observation.—When the absorbents of the abdomen, chest or head become inactive, or fail to take up the fluids secreted in their cavities, *dropsy* occurs in those parts. When the quantity of waste matter in different parts of the body is greater than the absorbents can remove, *tumors* are formed.

141. Though much impeded by the external layer of the skin, absorption takes place to a considerable extent through this membrane, and the use of medicinal baths is based on this fact; shipwrecked sailors, destitute of fresh water, find that thirst is relieved by immersing the body in salt water. Life is sometimes supported for a time by immersing the patient in baths of milk or broth.

Observation 1.—In serous and synovial membranes the fluids poured out into the joint in rheumatism and other inflammations are absorbed. Absorption is shown by injections of a solution of morphia under the skin to relieve suffering from neuralgic pain, from severe operations, obstinate cough and other irritations.

2.—In cases of disease where no food is taken into the stomach life is maintained by the absorption of fat. In consumption even the muscles and more solid parts of the body are absorbed.

Give the special office of the veins. Speak of the absorbing power of different membranes. How effected? How is dropsy produced? Tumors formed?

80 ANATOMY, PHYSIOLOGY AND HYGIENE.

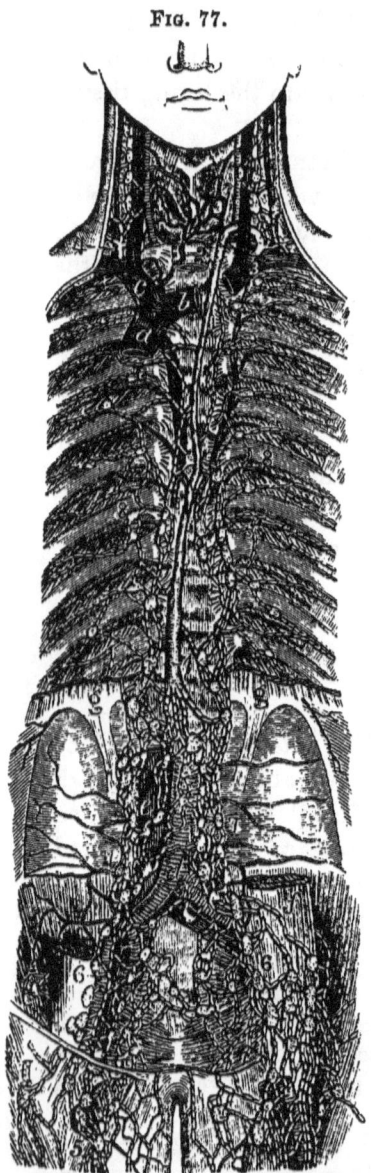

FIG. 77. VIEW OF THE GREAT LYMPHATIC TRUNKS.—1, 2, Thoracic duct. 4, The right lymphatic duct. 5, Lymphatics of the thigh. 6, Iliac lymphatics. 7, Lumbar lymphatics. 8, Intercostal lymphatics. *a*, Descending cava. *b*, Left innominate vein. *c*, Right innominate vein. *d*, Aorta. *e*, Ascending cava.

142. Absorption exists in the inferior animals. Those that live in a half torpid state during winter derive their nourishment from the absorption of the fat in different parts of their bodies. Swellings, bruises and ordinary injuries of animals are removed by absorption, as in Man.

§ **16.** HYGIENE OF THE ABSORBENTS.—*Conditions of Air affecting Absorption. Effect of Nutritious Food. Effect of the Removal of the External Layer of the Skin.*

143. *The air should be as free as possible from impure vapors and gases;* hence the importance of thorough ventilation, especially in the sleeping-room, since exhalations from the system are greater at night than by day.

Observation.—In infectious diseases the impure air should be constantly carried from the room, and the nurse should approach the patient *on the side* in which the currents of air are admitted.

144. *Moisture increases the activity of the absorbents;* hence, persons living in marshy districts contract miasmatic and contagious diseases more readily than those living in a drier atmosphere. In such local-

ities the house should be plentifully supplied with fresh air and kept dry by the use of fires. Especially is this necessary morning and evening in spring and autumn, and often in summer.

Observation.—For the above reason, the air of the sick room should be kept dry; otherwise the poisonous exhalations are absorbed by the lungs and skin both of the *patient* and of the *nurse*.

145. *Nutritious food lessens the activity of the absorbents;* hence, in cases of infectious diseases due attention should be given to the food of the attendants and of the family. Some persons use alcoholic stimulants or tobacco "to prevent taking disease," but these increase the activity of the absorbents and the liability to contract disease. A moderate amount of nutritious food will be more efficacious.

Observation 1.—*Absorption by the skin is most vigorous when the external layer is removed by blistering.* Then external applications, as ointments, are brought in immediate contact with the mouths of the lymphatics of the skin, and by them rapidly imbibed and circulated through the system. The same results follow if the skin is only punctured.

2.—*In handling poisons care should be taken that the external layer of the skin be unbroken,* as absorption is very rapid when it is removed. In contagious diseases, if the skin is broken it should be covered with adhesive plaster while at work over the patient. In handling dead bodies it is well to lubricate the hands with olive-oil or lard. The absorption of poisonous matter through a slight "scratch" or puncture of the cuticle, as the removal of a "hang-nail," has cost several valuable lives.

What should be the condition of the air? Observation. What influence has moisture? What care should be exercised by persons living in marshy districts? Observation. What is the influence of nutritious food upon absorption? Of alcoholic stimulants, etc.? When is absorption by the skin most vigorous?

SYNTHETIC TOPICAL REVIEW
OF THE ABSORBENTS.

State the Anatomy, the Physiology, the Hygiene of the Absorbents.

D *

CHAPTER VI.

THE RESPIRATORY AND VOCAL ORGANS.

146. IN previous chapters, we have noticed two stages in the change of food to form nutrient material and the transfer of the chyle into the vein which connects with the right chamber of the heart to mingle with the venous blood. The third and last change in Primary Assimilation is effected by the *Respiratory Organs*.

§ 17. ANATOMY OF THE RESPIRATORY AND VOCAL ORGANS.—*The Organs of the Voice. The Trachea—Bronchi—Lungs—Pleura. Blood-vessels of the Lungs. Diaphragm. Respiratory Muscles.*

147. The RESPIRATORY AND VOCAL ORGANS consist of the *Lar'ynx*, the *Tra'chea*, the *Bronch'i*, the *Lungs* and their *Blood-vessels*. The accessory organs are the *Thorax, Diaphragm* and *Respiratory Muscles*.

148. The LARYNX, the organ of the voice, is a short, cartilaginous cavity, extending from the root of the tongue to the trachea, with which it becomes continuous below. It is composed of five principal parts—the *Thy'roid*, the *Cri'coid*, the two *Aryte'noid* cartilages and the *Epiglot'tis*.

The THYROID is the largest cartilage. It consists of two wing-like plates, which meet in front and form the prominence called Adam's apple.

The CRICOID cartilage is about one-fourth of an inch wide in front and one inch behind.

The ARYTENOID cartilages are two in number, small, triangular and curved. They are placed upon the summit and back part of the cricoid cartilage, forming articulations.

The EPIGLOTTIS is oval-shaped, having its convex or bilg-

Of what do the Respiratory and Vocal organs consist? Describe the Larynx. Of what is it composed? What is said of the Thyroid cartilage? Of the Cricoid? Of the Arytenoid? Of the Epiglottis?

THE RESPIRATORY AND VOCAL ORGANS. 83

ing surface toward the mouth. It stands in a vertical position above the opening of the larynx, which is closed by it in the act of swallowing. (Figs. 78, 79.)

In the cavity of the larynx the mucous membrane is reflected at each side, outward and upward, forming a pair of pouches, called the *ventricles* of the larynx. Just below these ventricles are the true *vocal cords*, extending from a small

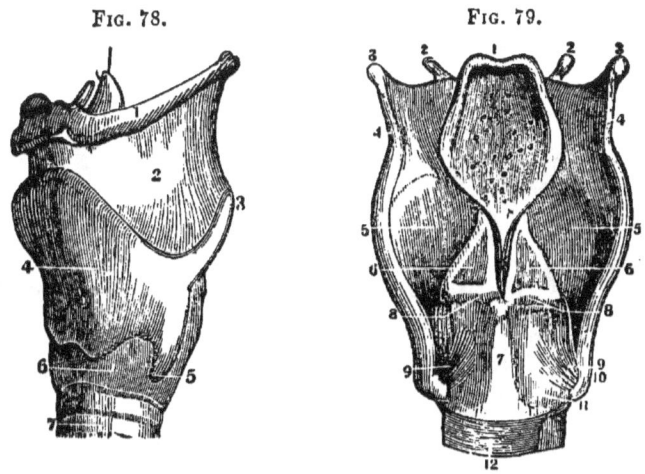

Fig. 78. Fig. 79.

Fig. 78. A SIDE VIEW OF THE CARTILAGES OF THE LARYNX.—*, The front side of the thyroid cartilage. 1, The os hyoides (bone at the base of the tongue). 2, The ligament that connects the hyoid bone and thyroid cartilage. 3, 4, 5, The thyroid cartilage. 6, The cricoid cartilage. 7, The trachea.

Fig. 79. A BACK VIEW OF THE CARTILAGES AND LIGAMENTS OF THE LARYNX.—1, The posterior face of the epiglottis. 3, 3, The os hyoides. 4, 4, The lateral ligaments which connect the os hyoides and thyroid cartilage. 5, 5, The posterior face of the thyroid cartilage. 6, 6, The arytenoid cartilages. 7, The cricoid cartilage. 8, 8, The junction of the cricoid and the arytenoid cartilages. 12, The first ring of the trachea.

process on the fore part of each Arytenoid cartilage to the recessed part of the Thyroid cartilage.

149. The TRACHEA (windpipe) is a vertical tube about an inch in diameter and four inches in length. It is made up of cartilage, muscle, and lined with mucous membrane. The Trachea is continuous with the larynx, and extends to the third dorsal vertebra, where it divides into two branches or tubes called *Bronchi*. (Fig. 81.)

What is the Trachea? What is the name of its two branches?

150. The Bronchi* are constructed like the Trachea. These tubes carry air to their respective lungs and again divide, sending a branch to each lobe. These divisions are repeated again and again until each ultimate ramification terminates in a group of small cavities called *air-cells*. (Fig. 81.)

151. The Lungs, consisting of two divisions, are situated in the cavity of the chest, enclosing between them the heart and the great blood-vessels. They accurately fill the cavity, adapting themselves to the varying size attending respiration.

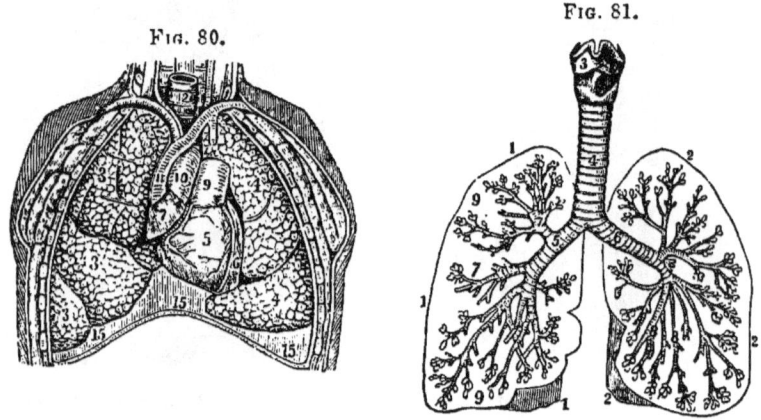

Fig. 80. The Lungs.—3, 3, 3, The lobes of the right lung. 4, 4, The lobes of the left lung. 5, 6, 7, The heart. 9, 10, 11, The large blood-vessels. 12, The trachea. 15, 15, 15, The diaphragm.

Fig. 81. The Bronchiæ.—1, Outline of the right lung. 2, Outline of left lung. 3, 4, Larynx and trachea. 5, 6, 7, 8, Bronchial tubes. 9, 9, Air-cells.

They are made up of numerous small lobules or clusters of air-cells, which unite into larger lobules. The lungs are closely invested with a membrane named *pleura*.

152. The Pleura is a serous membrane which lines the chest and then is reflected from the root of each lung over its surface.

Observation.—The lobules that compose the lungs seem to have no communication with each other, each lobule being in itself a miniature

---

Give the divisions and subdivisions of the Bronchi. Of how many divisions do the Lungs consist, and where situated? What is the Pleura? Observation.

* Gr., *brogchia*, the windpipe or throat.

THE RESPIRATORY AND VOCAL ORGANS. 85

lung, performing independent functions. It has been calculated that no less than eighteen thousand of these air-cells group around each terminal bronchial tube, giving a sum total of not less than six hundred millions. The lungs are everywhere unattached, excepting at the root, where they are firmly secured by the pulmonary ligaments, the pulmonary artery, the pulmonary veins and nerves and the bronchial tubes.

153. The BLOOD-VESSELS of the Lungs are the *Pulmonary* Artery,* the *Pulmonary Veins*, with the very minute hair-like vessels called *Capillaries.* (p. 222.)

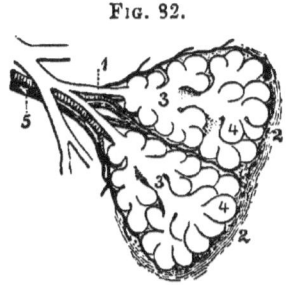

FIG. 82.

The PULMONARY ARTERY arises from the right ventricle of the heart and is distributed to the lungs.

The PULMONARY VEINS are four in number, two for each lung. They commence with the capillaries of the lungs and converge till a single trunk is formed for each lobe, or three trunks for the right lung and two for the left. The venous blood is conveyed to the lungs through this tube.

FIG. 82. DIAGRAM OF TWO PRIMARY LOBULES OF THE LUNGS, magnified.—1, Bronchial tube. 2, A pair of primary lobules connected by elastic tissue. 3, 3, 3, Inter-cellular air-passages. 4, 4, 4, Air-cells. 5, Branches of the pulmonary artery and vein.

The lungs, like other portions of the body, are supplied with nutrient blood-vessels.

154. The DIAPHRAGM is a flexible circular partition that separates the chest from the abdomen and the respiratory from the digestive organs. Its margin is attached to the spinal column, the sternum and cartilages of the lower ribs. The lungs rest upon its upper surface, while the liver and stomach are placed below it. In a state of repose its upper surface forms an arch, the convexity of which is toward the chest, or thorax (p. 19).

155. The RESPIRATORY muscles are in general attached at one extremity to the parts about the shoulders, head and upper portion of the spinal column. From these they run

---

Name the Blood-vessels of the Lungs. Speak of the Pulmonary Artery. Pulmonary Veins. Describe the Diaphragm. The Respiratory Muscles.

* Lat., *pulmo,* lungs.

downward and forward, and are attached at the opposite extremity to the sternum, clavicle and upper rib. Other muscles are attached at one extremity to a rib above and by the opposite extremity to a rib below. These fill the spaces between the ribs, and from their situation are called *intercost'al* muscles.

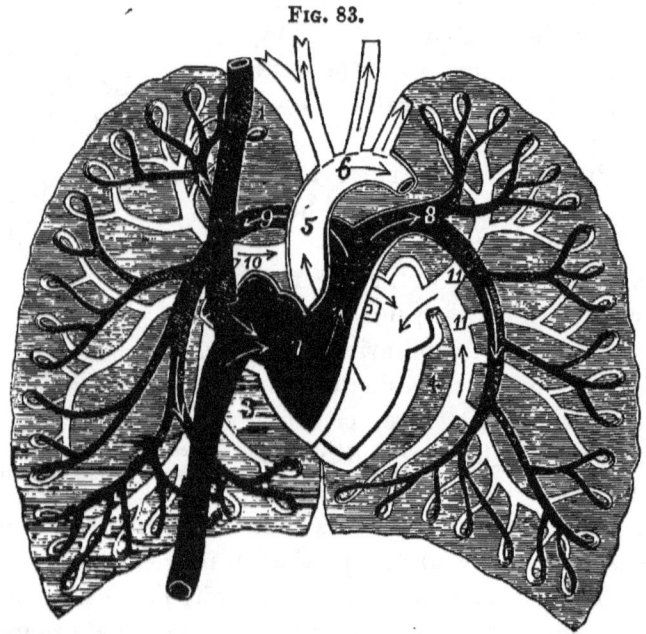

FIG. 83. PULMONARY CIRCULATION.—A DIAGRAM.—1, Descending vena cava. 2, Ascending cava vein. 3, Chamber of right side of heart. 4, Chamber of left side of heart. 5, Aorta. 6, Arch of aorta. 7, Pulmonary artery. 8, Its left branch. 9, Its right branch. 10, Right pulmonary vein. 11, 11, Left pulmonary vein.

§ 18. PHYSIOLOGY OF THE RESPIRATORY AND VOCAL ORGANS.—*The Function of Respiration. Two Modes of Respiration. Oxygenation of the Blood. Animal Heat. The Amount of Air in each Respiration. Conditions affecting the Number of Respirations. Double Function of the Larynx.*

156. We have traced the elaborated nutrient material to the blood. The Function of Respiration or breathing is to supply the blood with oxygen by *Inspiration,* and to remove from the blood carbonic acid by *Expiration.*

What is the function of Respiration?

157. In *Inspiration* the air is drawn into the lungs through the trachea, the muscular margin of the diaphragm contracts, which depresses its central portion; the chest is then enlarged, at the expense of the abdomen. At the same time that the diaphragm is depressed, the ribs are thrust forward and upward by means of muscles placed between and on them. Thus the chest is enlarged in every direction.

The lungs follow the variations of capacity in the chest, expanding their air-cells when the latter is enlarged and contracting when the chest is diminished. Thus, when the chest is expanded, the lungs follow, and consequently a vacuum is produced in their air-cells. The air then rushes through the mouth and nose into the trachea and its branches, and fills the vacuum as fast as it is made.

158. In *Expiration*, after the expansion of the chest, the muscles that elevated the ribs relax, together with the diaphragm. The elasticity of the cartilages of the ribs depresses them, and the cavity of the chest is diminished, attended by the expulsion of a portion of the air from the lungs. At the same time, the muscles that form the front walls of the abdominal cavity contract and press the alimentary canal, stomach and liver, upward against the diaphragm; this, being relaxed, yields to the pressure, rises upward and presses upon the lungs, which retreat before it, and another portion of air is expelled from these organs. These movements are successive during life, and constitute *Respiration*.

Observation.—The source of *Oxygen* is the air. It is everywhere ready to produce some new change in the materials of the organic and inorganic worlds. This life-giving principle does not exist free in the atmosphere, but is combined with *Nitrogen*, which is not a vital gas. The sources of *Carbonic* Acid (a poisonous gas) in animals are the blood and the tissues.

159. The two modes of RESPIRATION supply the blood with oxygen. The venous blood coming from all parts of the body has mingled with it, the elaborated chyle from

---

How is inspiration effected? What is said of the movements in expiration? State the sources of oxygen and carbonic acid.

the lacteals, and the lymph of the lymphatic vessels. This dark colored blood passes from the right side of the heart through the Pulmonary Artery and its capillaries in the lungs. These minute vessels interlace with and among the

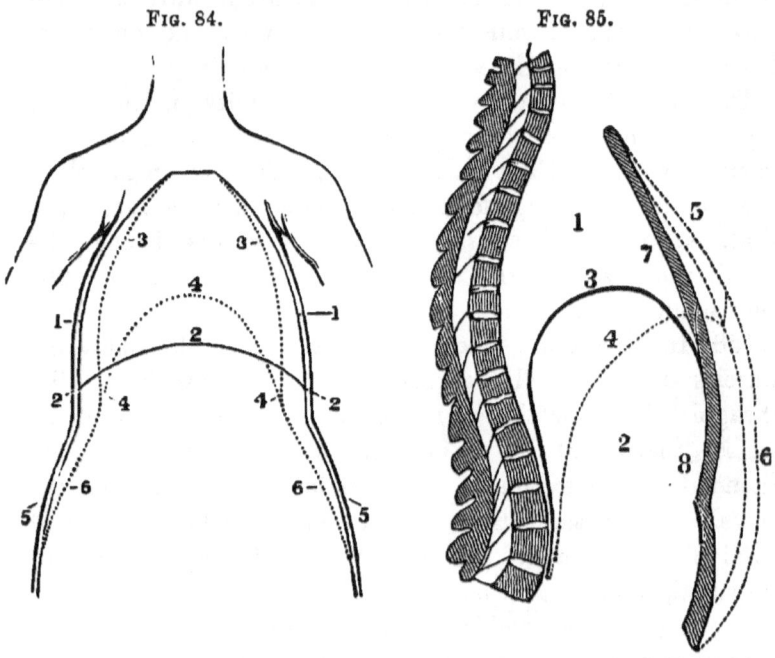

FIG. 84. A FRONT VIEW OF THE CHEST AND ABDOMEN IN RESPIRATION.—1, 1, The position of the walls of the chest in inspiration. 2, 2, 2, The position of the diaphragm in inspiration. 3, 3, The position of the walls of the chest in expiration. 4, 4, 4, The position of the diaphragm in expiration. 5, 5, The position of the walls of the abdomen in inspiration. 6, 6, The position of the abdominal walls in expiration.

FIG. 85. A SIDE VIEW OF THE CHEST AND ABDOMEN IN RESPIRATION.—1, The cavity of the chest. 2, The cavity of the abdomen. 3, The line of direction for the diaphragm when relaxed in expiration. 4, The line of direction for the diaphragm when contracted in inspiration. 5, 6, The position of the front walls of the chest and abdomen in inspiration. 7, 8, The position of the front walls of the abdomen and chest in expiration.

air-cells of the lungs. The inhaled air in these cells parts with oxygen and receives from the venous blood carbonic acid. This gas with vapor is expelled from the lungs through the trachea at every expiration. (Figs. 84, 85.)

Observation.—The presence of carbonic acid and watery vapor in the expired air can be proved by the following experiment: Breathe into lime-water, and in a few minutes it will become of a milk-white color. This is owing to the carbonic acid of the breath uniting with the lime, forming the *carbonate of lime*.

160. The oxygenation of the blood in the capillaries of the lungs changes its color from a dark maroon to a bright red. This purified or arterial blood is carried to the left side of the heart by the Pulmonary Veins, and it is now fitted for production of heat, motion and nervous energy.

Observation.—The venous blood holds in solution a considerable amount of carbonic acid, a little nitrogen, and a trace of oxygen. The exchange of oxygen and carbonic acid in the capillaries is effected partly by physical and partly by chemical processes. The carbonic acid exhaled nearly equals the amount of oxygen absorbed; the nitrogen exhaled and absorbed are nearly equal, and in addition there is the vapor or *pulmonary transpiration*.

161. The chemical changes in every part of the body caused by the union of oxygen with carbon, hydrogen and other elements of the blood and tissues, maintain the temperature of the body, and are the source of its nervous power and electricity.

The heat of the body, often called *Animal Heat*, is the result of the various chemical actions. The temperature of the tissues generally ranges from 98° to 100°; that of blood from 100° to 102°. The blood varies in temperature in different parts, being warmest in the hepatic veins of the liver.

Observation 1.—The amount of air taken in and given out in a respiratory movement must vary with different individuals and different conditions of the system. The volume of air ordinarily received by the lungs in a single inspiration is about one pint; the volume expelled, a little less than a pint.

2.—Respiration is more frequent in women and children than in men. Persons of small stature breathe more frequently but less deeply than taller people. In health, the smallest number of respirations in a minute by an adult is not less than fourteen, and they rarely exceed twenty-five; eighteen may be considered the average number. The number of respirations is increased by exercise, food, stimulants and moderate

---

State the exchange of oxygen and carbonic acid in the capillaries. Give observation. What is the color of the blood in the Pulmonary veins? For what is the arterial blood fitted? What are the results of chemical changes? What is the temperature of the tissues and the blood? Give the amount of air in respiration.

cold, while it is diminished by inactivity, moderate heat, starvation and general weakening influences, especially mental depression.

162. The LARYNX performs a double function, one part being concerned with respiration, the other with the voice.

In inspiration the vocal cords separate, allowing the air to pass in freely; in expiration they relax. The larynx, however, is the special organ of the voice, sounds being produced by the vibratory action of the vocal cords. During ordinary, tranquil breathing, the cords are widely separated, the glottis being of triangular shape; but when a vocal sound is to be produced the arytenoid cartilages are said to

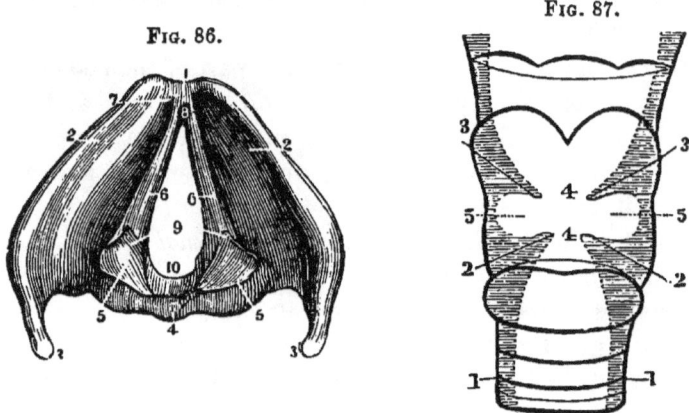

FIG. 86. A VIEW OF THE LARYNX, SHOWING THE VOCAL LIGAMENTS.—1, The anterior edge of the larynx. 4, The posterior face of the thyroid cartilage. 5, 5, The arytenoid cartilages. 6, 6, The vocal ligaments. 7, Their origin within the angle of the thyroid cartilage. 9, 10, The glottis.

FIG. 87.—AN IDEAL SECTION OF THE LARYNX.—1, The trachea. 2, 2, The lower vocal cords. 3, 3, The upper vocal cords. 4, 4, Glottis. 5, 5, The ventricles of the larynx.

become erect and almost to touch each other, the cords are made suddenly tense, closing the posterior portion of the glottis, while the anterior two-thirds opens a very fine fissure, and the air, driven by an unusually forcible expiration through the narrow opening, in passing between the vibrating vocal cord is itself thrown into vibrations which produce the sound required. (Figs. 86, 87.)

What is the office of the larynx in respiration? How are vocal sounds produced?

THE RESPIRATORY AND VOCAL ORGANS.

§ 19. HYGIENE OF THE RESPIRATORY AND VOCAL ORGANS.—*Importance of Proper Respiration. Composition of Pure Air—Of Respired Air. Importance of Ventilation in Public Rooms—In Sleeping Rooms—In Sick Rooms. The Effect of Compressing the Respiratory Organs: Means of Enlarging the Chest. The Influence of the Mind on Respiration. Beneficial Effects of Gymnastic and Calisthenic Exercises. Cultivation of the Voice.*

163. With minutest care, the hand Divine has arranged the unceasing play of the movements of the Respiratory Organs so that the air of the lungs is changed at every breath, giving us a requisite of health and even life—*pure blood.*

164. *Pure blood can be obtained only by a healthy action* of the respiratory organs, and this action only by a constant and sufficient supply of pure air. Limit this supply, and the stimulus furnished to the nervous and muscular tissues is withdrawn, and the carbonic acid is retained in the blood. Hence, the brain works sluggishly, the muscles become inactive, the heart acts imperfectly, the secretions are impaired, the food is not properly assimilated, and the whole body becomes enfeebled.

165. *Pure air* is composed of oxygen and nitrogen in about the proportion of 21 to 79. The air is most frequently rendered unfit for vital purposes by the presence of carbonic acid and minute particles of corrupted decaying animal matter. The amount of expired carbonic acid during the entire day is more than 15 cubic feet, or one pound and a half.

166. *The quality or purity of the air is affected by every respiration.* The quantity of nitrogen is nearly the same in the expired as in the inspired air, but the quantity of oxygen is diminished, and that of carbonic acid is increased. Thus, every time we force air from the lungs it becomes unfit to be breathed again.

Experiment.—Sink a "bell" jar that has a stop-cock into a pail of water, until the air is expelled from the jar. Fill the lungs with air, and retain it in the chest a short time, and then breathe into the jar

---

Why must there be a constant and sufficient supply of pure air? What is the composition of air? Why is exhaled air unfit to be breathed again? Experiment.

and instantly close the stop-cock. Close the opening of the jar that is under the water with a piece of paper laid on a plate of sufficient size to cover the opening, raise the jar from the water and invert it, then carefully sink into it a lighted candle suspended by a wire. The flame will be extinguished as quickly as if put in water. Remove the carbonic acid by inverting the jar, and place a lighted candle in it, and the flame will be as clear as when out of the jar.

Observation.—It is familiarly known that a taper will not burn where carbonic acid exists in any considerable quantity, or when there is a marked deficiency of oxygen. From this originated the judicious practice of sinking a lighted candle into a well or pit before descending into it. If the flame is extinguished, respiration cannot there be maintained, and life would be sacrificed should a person venture in until the impure air is removed.

167. *Air in which lamps will not burn with brilliancy is unfitted for respiration.* In crowded rooms, which are not ventilated, the air becomes foul, not only by a decrease of oxygen and an increase of carbonic acid, but by the waste, injurious atoms thrown out from the lungs and skin of the audience. The burning lamps, under such circumstances, emit but a feeble light. Let the oxygen gas be more and more expended, and the lamps will burn more and more feebly, until nearly extinguished.

Illustrations.—1. The effects of breathing the same air again and again are well illustrated by an incident that occurred in one of our colleges. A large audience had assembled in an ill-ventilated room to listen to a lecture; soon the lamps burned so dimly that the speaker and audience were nearly enveloped in darkness. The oppression, dizziness and faintness experienced by many of the audience induced them to leave, and in a few minutes after, the lamps were observed to rekindle, owing to the exchange of pure air, on opening the door, which supplied to them oxygen.

2. The "Black Hole of Calcutta" received its name from the fact that one hundred and forty-six Englishmen were shut up in a room eighteen feet square, with only two small windows on the same side to admit air. On opening this dungeon, ten hours after their imprisonment, only twenty-three were alive. The others had died from breathing impure air that contained animal matter from their own bodies.

168. *School-rooms, churches, concert-halls, and all rooms de-*

Observation. What remarks as to the necessity of ventilation of workshops, school-rooms, churches and concert-halls?

signed for public purposes and workshops should be amply ventilated. *The child at school becomes listless and uninterested ; why? Because he is stupefied by foul air.* When a pupil continues to breathe such air month after month, his brain is injured, and often consumption or other fatal disease destroys his young life, and then we wonder at the "mysterious providence" that takes from us the gifted and beautiful.

Observation.—The good man at church feels that he *ought* to be interested in the services, and yet, powerless to fix his attention, he sits nodding; why? *Because he is stupefied by foul air.* The air breathed over and over again last Sabbath and shut in during the week is all the poor man can obtain.

169. *The sleeping-room should be thoroughly ventilated.* Proper ventilation would often prevent morning headaches, want of appetite and general languor so common among the feeble. The impure air of sleeping-rooms probably causes more deaths than intemperance. Those who live in open houses little superior to the sheds that shelter the farmer's flocks are usually the most healthy and robust; headaches, liver complaints, coughs and a multitude of nervous affections are almost unknown to them. Not so with those who spend their days and nights in unventilated rooms with double windows, breathing over and over again the confined air; disease and suffering are their constant companions.

Observation.—Among children convulsions or fits often occur when they are sleeping, and not unfrequently in consequence of impure air. In such cases, by carrying the sufferer into the open air relief is afforded. Children should not sleep in *low beds* while adult persons occupy a higher bed in the same unventilated room, as carbonic acid is most abundant near the floor; nor is it advisable that the young sleep with the sick or aged.

170. *The ventilation of the sick room should receive special attention.* It is no unusual practice, when the patient is suffering from acute disease, as fevers, for the attendants to prevent the ingress of pure air, simply from fear that the

94 ANATOMY, PHYSIOLOGY AND HYGIENE.

sick person will take cold; and caution is indeed necessary, the patient should not *feel the current.*

Observation.—No room is suitable for sickness that is not so arranged that pure air may be constantly admitted without inconvenience or injury to the patient; and here we would say that *cool* air should not be mistaken for *pure* air. A very little sound judgment in this matter would doubtless save much suffering and lengthen life in a multitude of cases. The custom of having several persons sit in the sick-room vitiates the air and delays the recovery of the patient.

171. *The change that is effected in the blood while passing through the lungs, not only depends upon the purity of the air but the amount inspired.* The quantity varies according to the size of the chest, the movement of the ribs and diaphragm, and the health of the lungs. In children who have never worn close garments, the circumference of the chest is generally about equal to that of the body at the hips; and similar proportions would exist through life if there were no improper pressure of the clothing. Such is the case with the Indian woman, whose blanket allows the free expansion of the chest.

Observation 1.—The question is often asked, Can the size of the chest and the volume of the lungs be increased when they have been once compressed? Yes. The means to be used are, a full inflation of the lungs at each act of respiration, and a judicious exercise of them by walking in the open air, reading aloud, singing, sitting erect and practicing appropriate gymnastic exercises. Unless these exercises are systematic and persistent, they will not afford the beneficial results desired.

2.—Persons of sedentary habits should often, during the day, take full, deep breaths, filling the smallest air-cells with air; the shoulders should be thrown back and the head held erect.

172. *The size of the chest and lungs can be diminished by moderate and continued pressure.* This is most easily done in infancy, when the cartilages and ribs are very pliant; yet it can be effected at more advanced periods of life.

Observations.—1. The Chinese, by compressing the feet of female children, prevent their growth; so that the foot of a *Chinese belle* is not

---

What beside purity of air is required for proper respiration? Observations. How can **the size** of the chest be diminished?

THE RESPIRATORY AND VOCAL ORGANS. 95

larger than the foot of an American girl of five years. 2. The American women *compress their chests*, to prevent their growth; so that the chest of an *American belle* is not larger than the chest of a Chinese girl of five years. Which country, in this respect, exhibits the greater intelligence? 3. The chest can be deformed by making the linings of the waists of the dresses tight, as well as by corsets. Tight vests, upon the same principle, are also injurious.

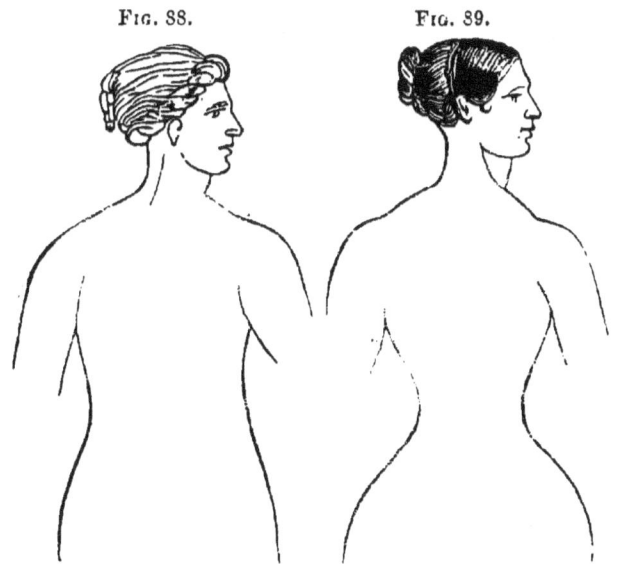

FIG. 88. FIG. 89.

FIG. 88. A CORRECT OUTLINE OF THE VENUS DE MEDICI, the *beau ideal* of female symmetry.
FIG. 89. AN OUTLINE OF A WELL-CORSETED MODERN BEAUTY.
One has an artificial, insect waist; the other, a natural waist. One has sloping shoulders, while the shoulders of the other are comparatively elevated, square and angular. The proportion of the corseted female below the waist is also a departure from the symmetry of nature.

173. *The position in standing and sitting influences the movement of the ribs and diaphragm.* When the shoulders are thrown back, and when a person stands or sits erect, the diaphragm and ribs have more freedom of motion, the abdominal muscles act more efficiently and the lungs have broader range of movement. When the lungs are properly filled with air, the chest is enlarged in every direction. If any article of apparel is worn so tight as to prevent the full ex-

---

State why position in standing and sitting influences respiration.

pansion of the chest and abdomen, the lungs, in consequence, do not receive air sufficient to purify the blood. The penalty for thus violating a law of our being is disease and suffering.

Observation.—But few persons realize the small amount of pressure that will prevent the enlargement of the chest. This can be shown by drawing a band tightly around the lower part of the chest of a vigorous adult and confining it with the thumb and finger. The restricted movements of the ribs are quite apparent in endeavoring to inflate the lungs.

174. *The state of the mind exercises a great influence upon respiration.* If we are depressed by grief or feel anxious about friends or property, the diaphragm and muscles that elevate the ribs will not contract with the same energy, the breathing is not as full or frequent as when the mind is influenced by joy, mirth and other enlivening emotions. Disappointed hopes is a frequent cause of consumption.

175. *Gymnastic and calisthenic exercises are invaluable aids to the culture and development of the voice,* and should be sedulously practiced when opportunity renders them accessible. A sedentary mode of life, the want of invigorating exercise, close and long-continued application of mind, and perhaps an impaired state of health or a feeble constitution, prevent, in many instances, the free and forcible use of those muscles on which voice is dependent.

176. *The art of cultivating the voice* has, in addition to the various forms of bodily exercise practiced for the general purpose of promoting health, its own specific prescription for securing the vigor of the vocal organs and modes of exercise adapted to the training of each class of organs separately. A few weeks of diligent cultivation are usually sufficient to produce such an effect on the vocal organs that persons who commence practice with a feeble and ineffective utterance attain in that short period the full command of clear, forcible and varied tone.

By what is respiration much influenced? Why should the student practice gymnastic and calisthenic exercises? What effect has culture on the voice?

§ 20. COMPARATIVE ANATOMY (Pneumonology).—*Respiratory Organs of Mammals—Of Birds—Of Reptiles—Of Amphibians—Of Fishes.*

177. The RESPIRATORY APPARATUS in other *Mammals* is similar to that of man both in structure and function. There are similar arrangements and movements of the ribs, sternum, intercostal muscles and diaphragm. The lungs fill the cavity of the chest, and have the same general composition of lobes, lobules and air-cells. (Fig. 90.)

Observation.—The development and health of domestic animals require good-sized lungs, unrestricted movement of the chest and pure air, as in man. An abundance of pure air is particularly needful to domestic fowls.

178. In *Birds* the lungs are confined to the back wall of the chest. They are not separated into lobes, but are oblong and flattened in shape, and connected with a series of air-receptacles scattered through various parts of the body. The ultimate pulmonary capillaries do not form a network lining definitely-bounded air-cells, as in mammals, but each vessel crosses an open-air space of its own. They interlace in every direction, forming a mass of capillaries permeated everywhere by air.

This arrangement not only reduces the specific gravity of the body, but also assists largely in the aeration of the blood. A marked modification of the respiration of birds of flight is the connection of the pores of the bones and feathers with the bronchial tubes and air-spaces of the lungs, so that there is an interchange of air between the lungs, the bones and the investing plumage. Birds consume more air in a given time proportionally than any other vertebrate, and they soonest die when deprived of it. (Figs. 91, 92.)

179. In *Reptiles* respiration is more simple than in mammals or birds. The lungs are less lobular and more bag-like, extending into the abdominal cavity. Upon the walls of these sac-like lungs the pulmonary vessels branch out. Owing

---

Comparative Pneumonology.—How does the Respiratory Apparatus in mammals compare with that in man? Describe the Lungs of Birds. What is said of the Ultimate Pulmonary Capillaries? What marked modification of respiration in birds of flight? Speak of respiration in Reptiles.

Fig. 90.

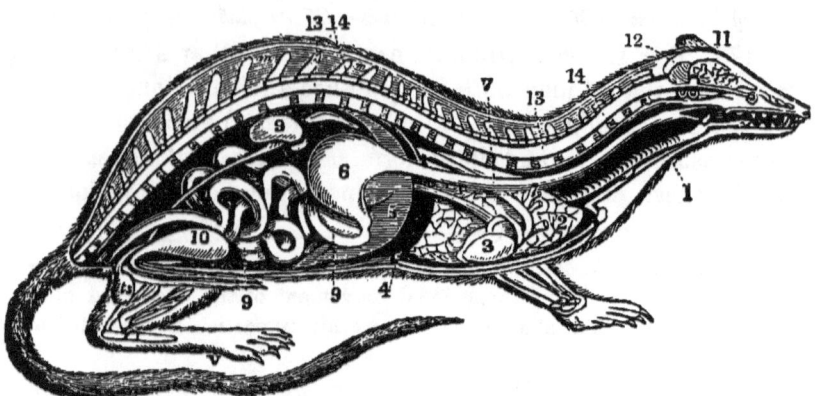

Fig. 90. Section of a Small Mammal.—1, Trachea. 2, Lungs. 3, Heart. 4, Diaphragm. 5, Liver. 6, Stomach. 7, Œsophagus. 8, Kidney. 9, 9, Intestines. 10, Bladder. 11, Cerebrum. 12, Cerebellum. 13, 13, Medulla spinalis. 14, 14, Vertebræ.

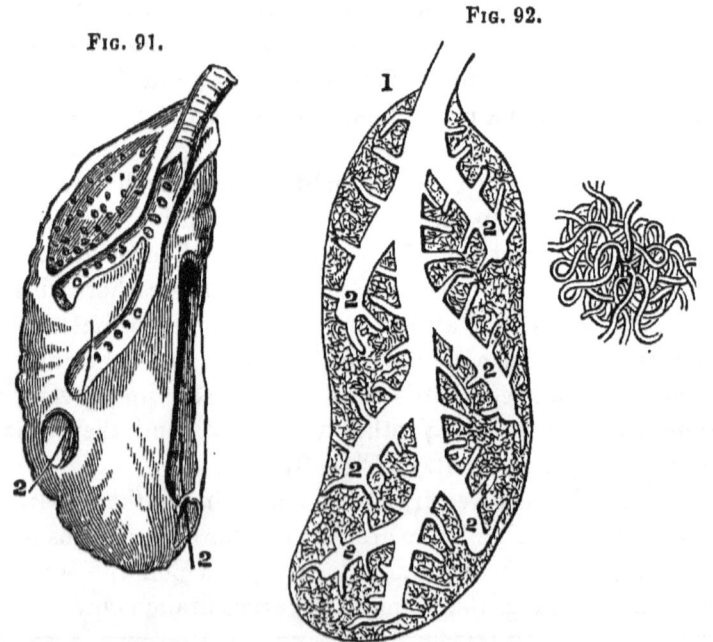

Fig. 91 (*Owen*). The Right Lung of a Goose.—1, A bronchial tube which divides into two tubes that open into the abdominal air-receptacles at 2, 2.

Fig. 92 (*Owen*). Ideal Section of a Bird, Magnified Two Hundred and Sixty Times.— 1, A bronchial tube that ends in a sac (cæca). 2, 2, 2, 2, 2, 2, Divisions of the bronchial tubes that branch out among the lobules. B, A plexus of capillary vessels.

THE RESPIRATORY AND VOCAL ORGANS. 99

to a less energetic respiration the movements of Reptiles are not so well sustained.

180. The AMPHIBIANS when young (tadpoles) breathe by gills; before becoming adult they acquire lungs, but the respiration is comparatively inactive. In Frogs the chest is not formed so as to act like a suction-pump, and accordingly these animals swallow the air by a sort of deglutition. (Fig. 93.)

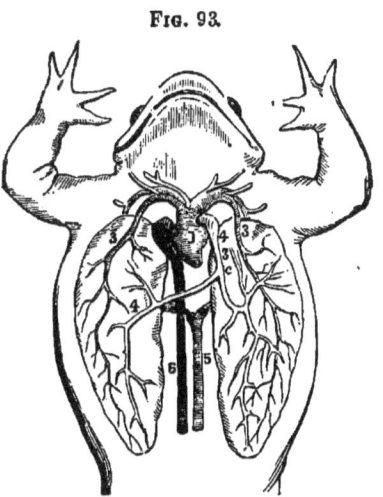

FIG. 93 (*Owen*). HEART AND LUNGS OF A FROG.—1, Heart. 2, Arch of the aorta. 3, 3, Pulmonary artery. 4, 4, Pulmonary veins. 5, 5, Aorta. 6, Vena cava.

181. In *Fishes* respiration is performed by means of the air dissolved in the water. Instead of lobular or bag-like lungs, there are found only a series of slit-like openings or arches on each side near the head, called the (branchiæ) gills. The bony and cartilaginous frames of these arches on the convex side support processes. On these are many plates or leaflets, covered by a delicate membrane (epithelium) on which the very minute capillary blood-vessels branch out. By this arrangement the blood-particles are more minutely separated and acted upon by the air in the water. In breathing, the mouth and gills of a fish open alternately; the water entering the mouth escapes by the openings of the gills. (Figs. 94, 95.) A remarkable feature in the organization of some fish is the swimming or air-bladder, placed in the abdomen under the back fin, communicating often with the œsophagus or stomach by a canal, permitting the escape of air from its interior. By a movement of the ribs the air-receptacle is acted on, so that by diminishing the quantity of air the specific

---

In Amphibians. In Fishes. What remarkable feature in the organization of some fish?

gravity of the fish alters according to circumstances. Fish that swim near the bottom have no air-bladder, as the Eel and Turbot. (Fig. 94.)

182. In some species of the *Annulosa*, as certain Spiders, the respiration is effected by air-bearing tubes (tracheæ), which

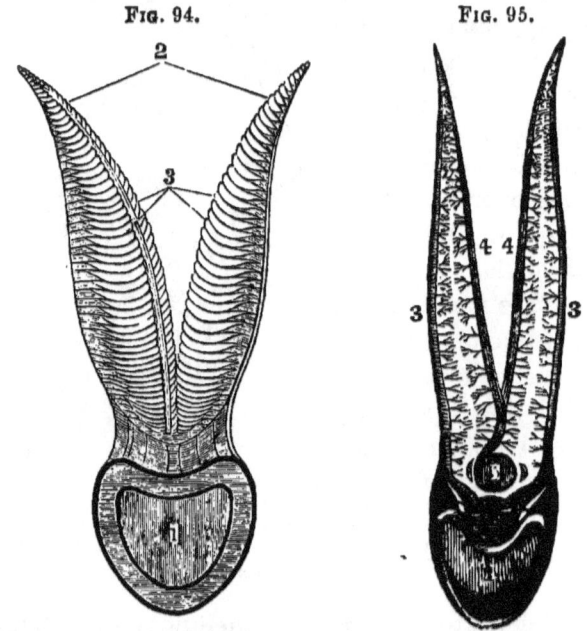

FIG. 94 (*Owen*). SECTION OF A BRANCHIAL ARCH, with a pair of processes supporting leaflets or plates from a cod, magnified two hundred and sixty diameters.—1, A section of a branchial arch. 2, A pair of processes. 3, Branchial leaflets or plates. The number of leaflets in one process of the cod is about one thousand; in the salmon fourteen hundred; in the sturgeon sixteen hundred.

FIG. 95 (*Owen*). A CIRCULATION OF THE BLOOD THROUGH THE BRANCHIAL LEAFLETS (a diagram).—1, A section of a branchial arch. 2, A section of a branchial artery. 3, An artery sent along the outer margin of the processes, giving off capillary vessels to the leaflets. 4, A vein that receives the blood from the capillaries on the inner margin of the process after the respiratory change has been effected and returns it to the branchial vein (5).

communicate with the exterior by small openings called *Stig'mata*. These openings often have valves which open and shut like the folding of a door. Through the air-bearing tubes the function of respiration is performed in every

---

Speak of respiration in some species of the Annulosa.

THE RESPIRATORY AND VOCAL ORGANS. 101

part of the body. This mode of breathing is peculiar to insects. (Fig. 96.)

183. In the *Mollusca* the respiratory organs vary. Some

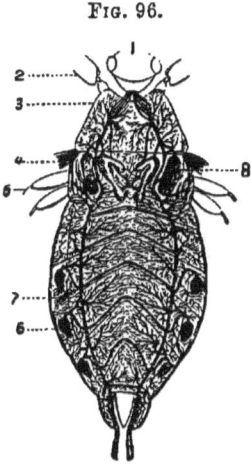

FIG. 96. THE RESPIRATORY ORGANS OF THE NEPA (Water Scorpion).—1, The head. 2, Base of the feet of the first pair. 3, First ring of the thorax. 4, Base of wings. 5, Base of the feet of the second pair. 6, Stigmata. 7, Tracheæ. 8, Aerial vesicles.

have the form of lungs, but in a state of great simplicity, like the Land-snail; while a large class have leaf-like gills, constituting what is known in the Oyster as the "beard."

Observation 1.—The Gills (branchiæ) of the Lobster are small bodies attached to the bases of the legs, and placed in a kind of chamber formed beneath the shield (carapace) on each side of the body. By the movements of the legs the water in the gill-chamber is constantly renewed.

2.—Some of the lowest class of animals are remarkable for the disposition of their respiratory apparatus, composed of membranous tubes spread out like a tree, as the Sea Cucumber.

Speak of respiration in Mollusca. In the Land-snail. In the Oyster. In the Lobster.

9*

102 ANATOMY, PHYSIOLOGY AND HYGIENE.

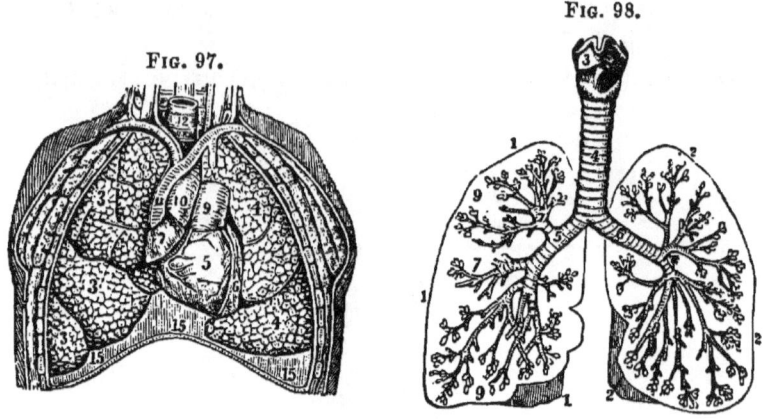

Fig. 97. Fig. 98.

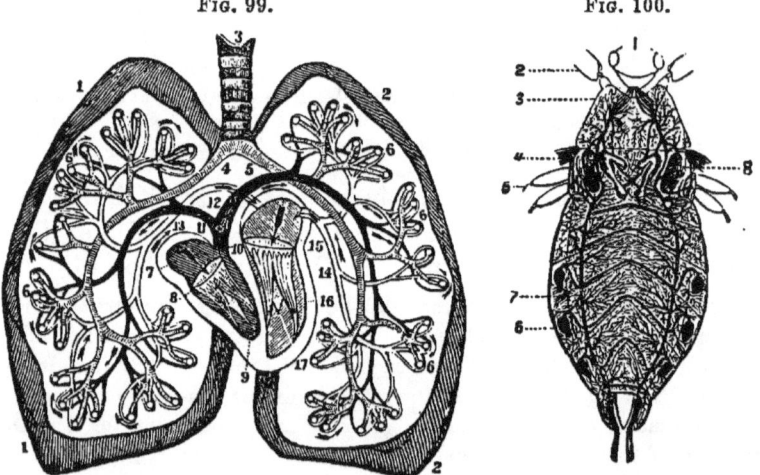

Fig. 99. Fig. 100.

SYNTHETIC TOPICAL REVIEW
OF RESPIRATORY SYSTEM.

Classes, Sub-kingdoms, Divisions, Anatomy, Physiology, Hygiene.

State the Anatomy, the Physiology and the Hygiene of the Respiratory System, Human and Comparative.

CHAPTER VII.

THE SKIN.

§ **21.** ANATOMY OF THE SKIN.—*The Skin an Eliminating Membrane. Layers of the Skin. The Epidermis. The Cuticle. The Dermis. The Papillary Layer. The Corium. The Hair-follicles. The Oil-glands. The Perspiratory Glands. Nails.*

184. WE have thus far spoken of the change in the blood by respiration; this bright red blood, freighted with nutrient material, is transmitted to the capillaries in all parts of the body. In these minute vessels another change is effected; and the useless matter is more or less expelled from the blood by the different vessels of the membrane that covers the body—the *Skin*.

185. The SKIN consists of two layers; a superficial one, destitute of nerves and blood-vessels, is called the *Epidermis*,* and a deeper layer, abundantly supplied with nerves and blood-vessels, called the *Dermis*, or *Cutis Vera* (true skin). (Fig. 101.)

186. The EPIDERMIS consists of two layers, different in many respects, one called the *Cuticle*, the other the *Soft Epidermis* (and named by some physiologists the Rete Mucosum). The epidermis holds the same relation to the dermis that the epithelium does to the deeper layer of the mucous membrane.

The CUTICLE is a horn-like membrane. Its deeper surface is continuous with the soft epidermic layer, from which it is constantly renewed. Its free surface is incessantly wearing away or shed in small flakes, constituting scurf or dandruff.

The SOFT EPIDERMIS consists of nucleated cells originating on the surface of the dermis. It is the seat of the color

---

Of what does the skin consist? Of what does the Epidermis consist? Give the relation of the Epidermis to the Dermis. What is the Cuticle? Of what does the Soft Epidermis consist? Where is the seat of color in the skin?

* Gr. *epi*, upon, and *derma*, skin.

of the skin. The difference between the blonde and the brunette, the European and the African, lies in this tissue.

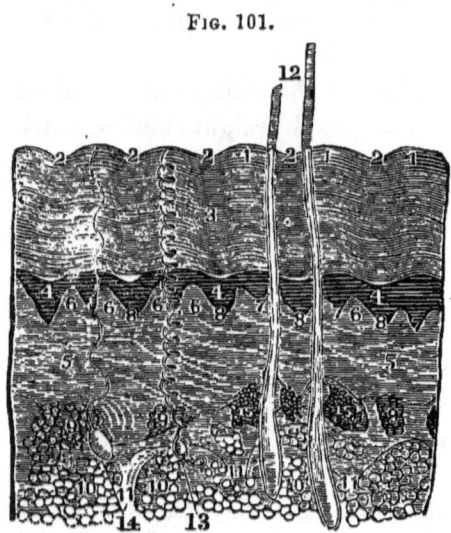

FIG. 101. A DIAGRAM OF THE SKIN.—1, 1, The lines or ridges of the cuticle, cut perpendicularly. 2, 2, 2, 2, 2, The furrows or wrinkles of the same. 3, The cuticle. 4, 4, The colored layer of the cuticle. 5, 5, The cutis vera. 6, 6, 6, 6, 6, The papillæ, each of which answers to the lines on the external surface of the skin. 7, 7, Small furrows between the papillæ. 8, 8, 8, 8, The deeper furrows between each couple of the papillæ. 9, 9, Cells filled with fat. 10, 10, 10, The adipose layer, with numerous fat vesicles. 12, Two hairs. 13, A perspiratory gland, with its spiral duct. 14, Another perspiratory gland, with a duct less spiral. 15, 15, Oil-glands with ducts opening into the sheath of the hair (12).

187. The DERMIS or True Skin presents two very different surfaces, of which the external is called the *Papillary* layer, the internal the *Corium*.

The PAPILLARY or outer layer of the dermis is provided with a multitude of little conical-shaped projections. These are prolongations of the upper compact tissue of the corium into the newly-formed layer of the epidermis. They vary in number and degree of development in different parts of the body. The papillæ are very numerous on the palm of the hand and on the free border of the lips.

The CORIUM is made up of interlacing bundles of fibres. These are so interwoven as to constitute a firm, flexible web. Here the arteries of the skin penetrate from beneath, and end in a capillary network, from which looped vessels project and enter the papillary layer. The veins emerging from the skin are more numerous and much larger than the arteries. The lymphatics also form a close network on the surface. The skin is abundantly

---

What is said of the Dermis? Describe the Papillary layer. Speak of the Corium. Of its vessels.

supplied with nerves, but their mode of termination has not been accurately ascertained.

188. Buried in the Corium are *Hair-follicles* or sacs. Each hair-follicle receives, in nearly all cases, the ducts of two *Sebaceous* or *Oil-Glands*, which are situated in the dermis. They are found only where hair exists. Each gland is a flask-shaped body, composed of from five to twenty little sacs, clustered around and leading into a common duct. These glands are lined by a fine epithelium, and the oily secretion first anoints the hair-bulb and then oozes out upon the neighboring surface of the cuticle.

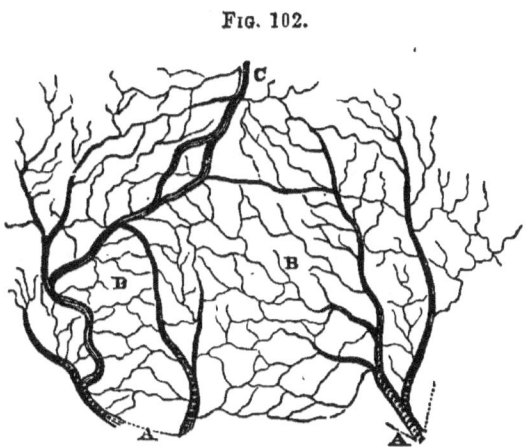

FIG. 102. THE ARTERIES AND VEINS OF A SECTION OF THE SKIN.— A, A, Arterial branches. B, B, Capillary, or hair-like vessels, in which the large branches terminate. C, the venous trunk, collecting the blood from the capillaries.

Observation.—At the bottom of the follicle is a more or less elevated portion of the dermis, often forming a distinct papilla, which is destitute of cuticle. The root of the hair is composed of soft, pale and somewhat compressed nucleated cells; it is adherent to the lining of the follicle or *root-sheath*. When a hair is plucked out, the sheath adheres to it, but the vascular papilla at the bottom of the follicle remains, and a new hair is generated upon it. If the papilla is destroyed, no new hair can be formed. All these papillæ, except those of the finest hairs, probably receive nervous fibrils. The part of the hair projecting above the surface is called the *shaft*.

189. Immediately beneath the skin, over the whole surface of the body, there are a multitude of little glandular bodies,

---

Describe the Hair-follicles. Describe the different parts of a hair. Describe the Oil-glands.

called *Perspiratory* or *Sweat Glands*. Each gland consists of a minute, cylindrical, spiral duct, which passes inward through the epidermis, and terminates in a globular coil in the deeper

FIG. 103.

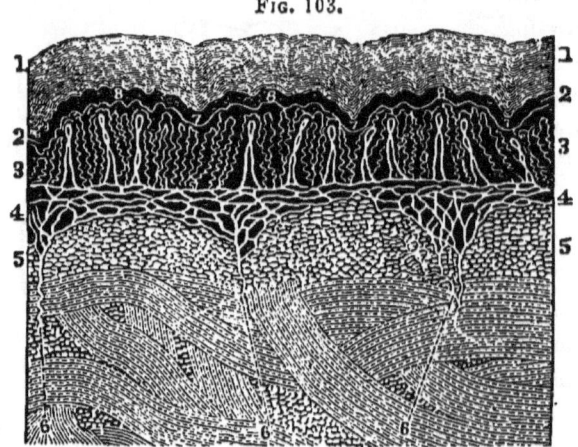

FIG. 103. A DIAGRAM.—1, 1, The cuticle. 2, 2, Its soft layer. 4, 4, The network of nerves. 5, 5, The Dermis. 6, 6, 6, Three nerves that divide from the network (4, 4).

meshes of the true skin. The opening of the duct upon the cuticle is called the "pore." (Fig. 101.)

Observation.—This aperture is oblique in direction, and possesses all the advantages of a valvular opening, preventing the ingress of foreign injurious substances to the interior of the duct or gland. It is estimated that six thousand glands exist on every square inch of surface, and the combined length of the glandular tubing in the body is between two and three miles. These glands, coming in contact with the capillary blood-vessels, receive a watery fluid (the perspiration) from the blood. The formation of perspiration is constant, but usually evaporation takes place as fast as it reaches the surface. This is called the "insensible transpiration" of the skin.

190. The NAILS are horny appendages of the skin, and correspond with the hoofs and claws of animals. They are flexible plates continuous with the epidermis, and rest on the depressed surface of the dermis, called the bed. By maceration or severe scalding, even in life, the nail is detached with the epidermis.

Where are the Sweat-glands? What are "pores"? What is "insensible transpiration"? Speak of the Nails. Of what is the horny part composed?

§ **22.** PHYSIOLOGY OF THE SKIN. *Functions of the Skin. Uses of the Epidermis. The Corium. The Lymphatic Vessels of the Skin. Function of the Oil-glands. Uses of Perspiration. Conditions that Modify Quantity of Perspiration. The Function of the Hair—Of the Nails.*

191. The FUNCTIONS OF THE SKIN are threefold: 1st, As a *Protecting* membrane; 2d, As a *Medium* for the distribution of the nerve-filaments of touch; and 3d, As an *Eliminating* or discharging organ.

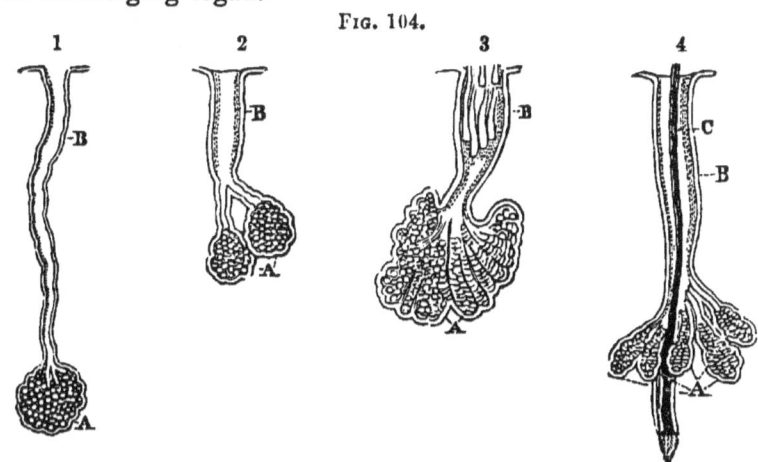

FIG. 104. OIL-GLANDS AND DUCTS, magnified thirty-eight diameters. 1, A, Oil-gland from the scalp; B, Its duct. 2, A, Two glands from the skin of the nose; B, Common duct. 3, A, Oil-gland from the nose; B, The duct filled with the peculiar animalculæ of the oily substance; the heads are directed inward. 4, A, Cluster of oil-glands around the shaft of the hair (C); B, Ducts.

192. The uses of the *Epidermis* are various. It serves to cover and protect the delicate sensitive parts beneath it; to prevent the too rapid escape of heat; and to restrain the evaporation of the fluids of the skin and its appendages, at the same time that it furnishes a medium through which these secretions can reach the surface of the body.

Observation.—The cuticle is constantly destroyed and replaced, as is proved by the disappearance from the skin of such stains as those produced by nitrate of silver, or the scales thrown off after some acute diseases, as scarlatina. The restoration of the cuticle is observed after the process of vesication by blisters, and in consequence of burns and scalds. By these means large patches of cuticle are removed; but they are re-

---

Name the functions of the Skin. What are the uses of the Epidermis?

newed in a short time, under favorable circumstances. The colored substance is also capable of rapid reproduction.

193. In the *Corium*, or internal layer of the skin, resides vitality. It varies in thickness in different parts of the body. The unevenness of this soft layer of the skin has reference to an important law in animal organization—that of multiplying surface for the increase of function. This object is effected by the little prominences or *Papillæ*.

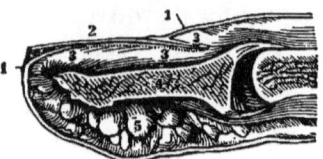

FIG. 105. A SECTION OF THE END OF THE FINGER AND NAIL.—4, Section of the last bone of the finger. 5, Fat, forming the cushion at the end of the finger. 2, The nail. 1, 1, The cuticle continued under and around the root of the nail at 3, 3, 3.

194. The surface of the skin possesses the power of absorbing both liquids and vapors. The principal, if not sole, agents of this function on the surface of the body are the cutaneous *Lymphatic* vessels, which are active in proportion to the thinness or absence of the cuticle. To a slight extent the skin is a respiratory membrane in man, giving off carbonic gas and actually absorbing oxygen.

195. The *Sebaceous matter* from the *Oil-glands* anoints the hairs with oil in their progress of growth from the skin, and also imbues or covers the cuticle, by which it is rendered repellant of water.

Observation.—The oiliness of the surface of the skin, occasioned by this material, permits the ready adhesion of dust and dirt, and necessitates the use of soap for the easy removal of its excess. This oily product often becomes thickened and distends the glands, most frequently in the face, and especially on the nose, and at the mouths of the ducts it becomes mixed with dust. When pressed out it assumes the spiral form of the duct; hence it is commonly taken for a worm. In the healthiest individuals the sebaceous matter contains a curious parasite, called the "pimple mite."

196. The uses of the *Perspiration* or sweat are twofold: 1st, To free the system of a certain quantity of water; in effecting this purpose, the perspiratory function becomes a

---

Describe the Corium. State the use of the Lymphatic vessels of the Skin. What are the uses of the Oil-glands? What are the functions of Perspiration?

regulator of the temperature of the body; 2d, To expel from the body certain special products of chemical changes. Sanctorius estimated that *five* of every *eight* pounds of food and drink was discharged from the body through the many outlets of the skin.

Observation 1.—The quantity of perspiration exhaled by different parts of the body differs widely. Its general quantity is influenced both by internal and external conditions; thus, it is increased by a higher temperature of the body, by a quicker circulation, and therefore by exercise and effort generally. Perspiration may also be induced by additional covering of the body, and also by peculiar conditions of the nervous system.

2.—Of the external conditions which modify the quantity of perspiration, the *condition of the atmosphere* is most important. Thus, in warm air the activity of the circulation of the skin is increased, which increases the perspiration, whilst cold air has the opposite effect; again, dry air increases the perspiration, whilst damp air diminishes it.

3.—The skin is said to regulate the quantity of fluid given off by the kidneys and the quantity of fluid left in reserve in the blood and soft tissues generally. Observation shows that in cold weather the skin exhales less and the kidneys excrete more fluid, while in warm weather the skin expels more and the kidneys less.

197. The use of the insensible outgrowth of the epidermis, the *Hair*, is protection; and the function of the *Nails* is not only protection, but support to the yielding softness of the flesh at the finger-tips.

§ **23.** HYGIENE OF THE SKIN. *The Condition of the Skin. Clothing. Kind of Material for Clothing. Class of Persons that need More Clothing. Cleanliness of Clothing. Bathing—Modes of Bathing—Time for Baths—General Rules for Bathing—Water a Curative Agent. Air Beneficial to the Skin. Effect of Light on the Skin.*

198. The skin is in constant activity; the watery part of the perspiration is removed by evaporation, the more solid waste matter is deposited upon the surface of this membrane. To maintain its healthy action in every part, attention must be given to *Clothing, Bathing, Light* and *Air*.

199. CLOTHING is chiefly useful in preventing the escape

Give the uses of the Hair—Of the Nails. What is said of the use of Clothing?

110 ANATOMY, PHYSIOLOGY AND HYGIENE.

of too much heat from the body, and in protecting the body from exposure to the evil effects of a varying temperature of the atmosphere. In selecting and applying clothing, the following should be observed:

200. *The material for clothing should be a bad conductor of heat.* As air is a non-conductor, material should be chosen which is capable of retaining much air in its meshes, and as moisture increases the conducting power, the material should not be such as will absorb or retain moisture.

Observation.—*Furs* retain much air in their meshes and absorb scarcely any moisture, and consequently are well adapted to those subject to the great exposures of very cold climates. *Woolen cloth*, next to furs and eider down, retains the most air and absorbs the least moisture; hence it is a good article of apparel for all persons, unless too irritable to an over-sensitive skin. In that case the flannel may be lined with cotton, or *silk* may be substituted. When of sufficient body or thickness, silk is a good article for inner clothing, excepting when it produces too much disturbance of the electricity of the system. Next to these articles, *cotton* is well adapted for garments worn next the skin. *Linen* should never be worn by persons in any way enfeebled, even in warm weather or in hot climates. It is a good conductor of heat and readily absorbs moisture; hence, with such covering, the body is surrounded by a layer of moisture instead of air.

201. *The clothing should be both porous and loosely fitted.* The necessity of porous clothing is seen in the wearing of India-rubber overshoes. In a short time the hose and under-boot become damp from retained perspiration. The waste matter thus left in contact with the skin is reconveyed into the system by absorption, causing headache and other diseases. Free exhalation, and a layer of air secured by loose clothing, enable the skin to absorb oxygen, which gives it tone and vigor.

Observation.—As the design of additional clothing is to enclose stratums or layers of warm air, we should, in going from a warm room into cold air, put on our extra covering some time previous to going out, that the layers of air which we carry with us may be warmed by the heat of the room, and not borrowed from the heat of the body.

202. *The clothing must be suited to the state of the atmosphere*

Of what material should it be? Observation. Why should the clothing be porous and loosely fitted? Observation. To what must it be suited?

and to the condition of the individual. Sudden changes of temperature should be regarded; but it is usually unsafe to make changes from thick to thin clothing, excepting in the morning, when the vital powers are in full play. The evening usually demands an extra garment, as the atmosphere is more cool and damp, and we have also less vital energy than in the early part of the day.

Observation.—Many a young lady has laid the foundation of a fatal disease by exchanging the thick dress, warm hose and shoes for the flimsy fabric, thin hose and shoes which are considered suitable for the ball-room or party. All sudden changes of this kind are attended with hazard, which is proportionate to the weakness or exhaustion of the system when the change is made.

203. *The child and the aged person require more clothing than the vigorous person of middle age.* Judging from observation, we should infer that children needed less clothing than adults. The exposure to which the vain and thoughtless mother subjects her child very frequently lays the foundation for future disease. Those who have outlived the energies of adult life also need special care regarding a proper amount of clothing.

Observation.—The system of "hardening" children, of which we sometimes hear, is as inhuman as it is unprofitable. To make the child robust and active, he must have nutritious food at stated hours, free exercise in the open air, and be guarded from the cold by *proper apparel.*

204. *When a vital organ is diseased, more clothing is needed.* In consumption, dyspepsia, and even headache, the skin usually is pale and the extremities cold, because less heat is generated. Persons suffering from these complaints need more clothing than those with healthy organs.

205. *Persons of active habits need less clothing than those of sedentary employment.* Exercise increases the circulation of the blood, consequently the vital activities become more energetic, and more heat is produced. We need less clothing when walking than when riding.

Observation. Who require the more clothing? Observation. What is said of clothing when a vital organ is diseased? What persons need less clothing? What is said of cleanliness of the clothing?

206. *The clothing should be kept clean.* Some portion of the exhaled fluids of the body must necessarily be absorbed by the clothing; hence, warmth, cleanliness and health require that it should be frequently changed and thoroughly washed. Under-garments worn through the day should not be worn through the night, nor the reverse. When taken from the body, such garments should not be hung in the closet or put into the drawer, but exposed to a current of fresh air.

207. *Damp clothing is injurious.* All articles from the laundry should be well aired before being worn. When the clothing is wet by accident or exposure, it should be changed immediately, unless the person is exercising so vigorously as to prevent the slightest chill. When the exercise ceases, the body should be rubbed with a dry crash towel till a thorough reaction takes place.

Observation.—The covering of beds should be thoroughly aired every morning, and frequently renewed. Beds and bedding that have not been used for some weeks become damp, and should be dried before use. A hostess cannot be guilty of a more inhospitable act than that of sending her guest to her fine guest-chamber, to occupy a bed which has been long unused.

208. BATHING is indispensable to sound health as well as to cleanliness. The skin soon becomes covered with a mixture of perspirable matter, oil and dust, which, if allowed to remain, interferes with the action of the skin as an excretory organ. This increases the action of the lungs, kidneys, liver, etc., which take upon themselves the excretory work which the skin fails to perform.

Observation.—By overwork the liver becomes diseased, and if it is continued, the result will be consumption and other diseases of the vital organs. Again, obstruction of the pores will prevent respiration through the skin, and deprive the blood of one source of its oxygen and one outlet of its carbonic acid.

209. *Bathing gives tone and vigor to the internal organs.* When cool water is applied to the body, the skin instantly

---

What of damp clothing? What is indispensable to health? What effect has bathing on the internal organs? Give observation.

shrinks and the whole of its tissue contracts. This contraction diminishes the capacity of the blood-vessels, and a portion of the blood is thrown upon the internal organs. The nervous system is stimulated, and communicates its stimulus to the whole system. This causes a more energetic action of the heart and blood-vessels, and a consequent rush of blood back to the skin. This is the state termed *reaction*, the first object and purpose of every form of bathing.

Observation 1.—By this reaction the internal organs are relieved, respiration is lightened, the heart is made to beat calm and free, the tone of the muscular system is increased, the appetite is sharpened, the mind more clear and strong, and the whole system seems to possess new power. Regularity in bathing is necessary to produce permanently good effects.

2.—The simplest modes of bathing are by means of the sponge or the shallow baths. The body may be quickly sponged over, wiped dry and followed by friction. The water may be warm or cold. If cold, the bath should be taken in the early part of the day, and followed by exercise. If exercise cannot be taken, the individual should rest under covering. The warm bath should usually be taken just before retiring.

210. *The frequency of bathing must depend upon the condition and occupation of the individual.* Daily bathing may be practiced with profit by most persons, but to the studious and sedentary it is in most cases absolutely indispensable.

The hour for ablution is of importance. It should neither immediately precede nor follow a meal. The same is true of severe mental and muscular exercise. The bath is less beneficial in the afternoon than the forenoon. The best time for cold baths is two or three hours after breakfast. The system is then at "flood-tide," while from that time till the retiring hour the tide is ebbing; hence, the worst time for a cold bath is at bed-time.

Observation 1.—For those who cannot choose their time, the hour of rising will answer very well—that is, for many persons, especially if they become accustomed to the use of water by beginning at another and a better hour. If the mind and body are brightened by the early

---

Upon what must depend the frequency of bathing? What should the time be?

bath, and an exhilaration follows, the bath is beneficial; if, on the contrary, languor follows, and the skin looks blue or too pale, it is injurious. That the bath is to be followed by exercise must not be forgotten.

2.—In diseases of the skin, and many chronic ailments of the internal organs, bathing is a remedial measure of great power. In disease which has baffled the skill of physicians depending wholly upon internal remedies, the effect of a systematic course of baths is often surprising. Like other curative means, the baths should be directed by those who thoroughly understand the use of water as a remedial agency. Matters of diet, exercise, etc., require adaptation to the treatment of the particular case. Those who desire the *full benefit* of these means must avail themselves of the appliances of a well-conducted hygienic establishment.

211. Pure AIR is an agent of great importance in the functions of the skin. It imparts to this membrane some oxygen, and receives from it carbonic acid gas. It likewise removes perspiration and portions of the oily secretion.

212. LIGHT exercises a very salutary influence upon the skin. It is no less essential to the vigor of animal than of vegetable life. Dwelling-houses should be built with reference to the free admission of sunlight and air into all occupied rooms.

Observation.—The dark, damp rooms so much used by indigent families and domestics in cities and large villages are fruitful causes of vice, poverty and suffering. Ladies often suffer seriously from too much exclusion of sunlight. Excepting in very warm weather, they should practice sitting or exercising in the full sunshine of the out-door world.

In what diseases is bathing of great importance? Observations. State the influence of pure air. What influence does light exercise? Give a cause of vice.

SYNTHETIC TOPICAL REVIEW.

EPIDERMIS—DERMIS—VESSELS OF THE SKIN—APPENDAGES OF THE SKIN.

State the Anatomy, the Physiology, the Hygiene of the Skin and its Appendages.

CHAPTER VIII.

THE CIRCULATORY ORGANS.

213. THE *Blood* is the most important as well as the most abundant fluid in the body. When expelled from the lungs to the left cavity of the heart, it contains all the materials for the support of every part of the animal frame. In order that the blood with its cargo of supplies should fulfill its mission of nutrition, it must be kept constantly moving in a circuit, or from the heart outward to every part of the body, and from the tissues backward to the heart to be renewed. This movement is called its *Circulation*. The organs through which the blood circulates are the *Heart*, the *Arteries*, the *Veins* and the *Capillaries*.

§ **24.** ANATOMY OF THE CIRCULATORY ORGANS.—*Construction of the Heart. The Arteries, Veins and Capillaries, and their Relation to each other. The Aorta and its Divisions. Arrangement of the Veins.*

214. The HEART is placed in the left side of the chest, between the right and left lung. It is a hollow muscle enclosed in a sac, named *Pericardium*. Its length is about five inches, and its basal diameter about four inches. The heart consists of four cavities or chambers, two on each side, called *Auricles* and *Ventricles*. The auricles receive the blood coming into the heart by the *veins*, the ventricles expel the blood out of this organ by the *arteries*. (Fig. 107.)

215. The PERICARDIUM (the heart case) is a membrane not only spread over the external surface of the heart, but is reflected or doubled in itself, so as to form a loose sac; it protects the heart from friction against other parts. (Fig. 106.)

216. The AURICLES differ in muscularity from the ventri-

What is said of the blood? Why must the blood be kept in circulation? Name the Circulatory organs. Describe the heart. The Pericardium. How many cavities? What are they called? Describe the Auricles.

cles. Their walls are thinner, and of a bluish color. They occupy the basal end of the heart.

217. The VENTRICLES not only have thicker walls than the auricles, but those of the left ventricle are thicker and stronger than the right. In the interior of these cavities arise fleshy columns called *colum'næ car'neæ.*

218. Folds of membrane called VALVES are found between the auricles and ventricles. In the right side of the heart there are three folds or doublings, called the *tri-cus'pid* valves. Between the auricle and ventricle, in the left side, there are two valves, called the *mi'tral.* There are seen passing from the floating edge of these valves to the columnæ carneæ,

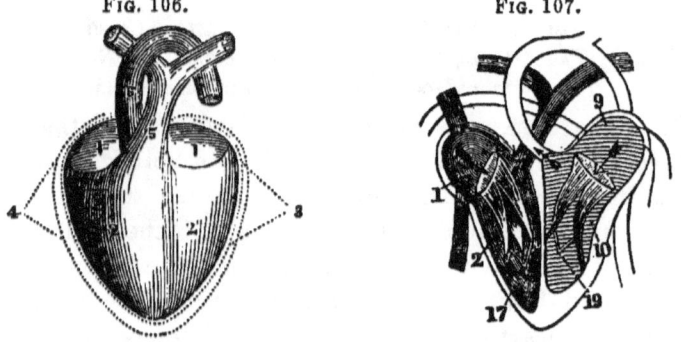

FIG. 106. A DIAGRAM OF THE HEART.—1, 1, Right and left auricle. 2, 2, Right and left ventricle. 3, 4, The pericardium. 5, 6, Large blood-vessels (arteries).

FIG. 107. IDEAL HEART.—1, Right auricle. 2, Right ventricle. 9, Left auricle. 10, Left ventricle. 17, Tri-cuspid valves. 19, Mitral valves.

small white cords, called *chor'dæ ten'di-næ*, which prevent the floating edge of the valve from being carried into the auricle. (Fig. 107.)

219. The ARTERIES are firm, membranous tubes, arising from the ventricles of the heart by two trunks; that from the left ventricle, named the *Aorta*, is the systemic trunk: and that from the right ventricle, named the *Pulmonic artery*, is the pulmonic trunk. (Fig. 108.)

220. The AORTA rises from the left ventricle for a short

Describe the Ventricles. The Valves of the heart. What are the Arteries? From what part of the heart arises the Aorta, and give its course?

THE CIRCULATORY ORGANS.

Fig. 108.

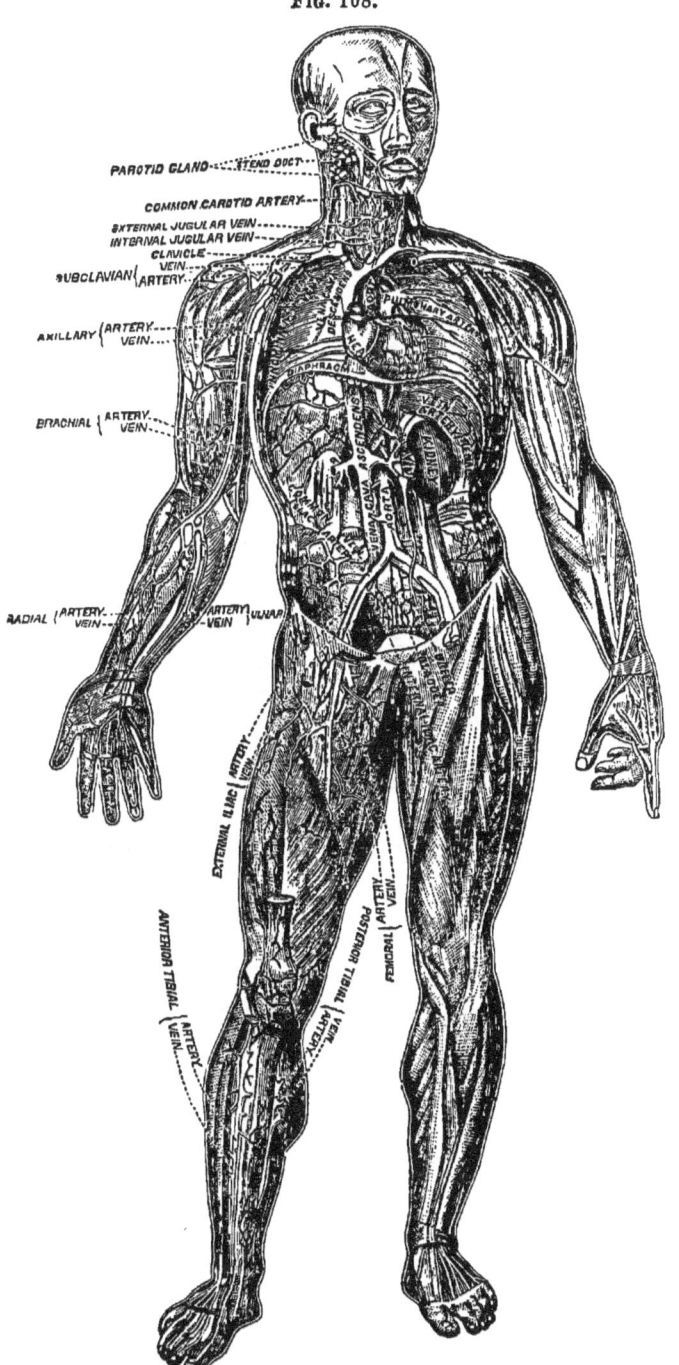

118   ANATOMY, PHYSIOLOGY AND HYGIENE.

distance behind the sternum and then curves downward, forming a semicircular bend, called the *Arch* of the Aorta. It then passes downward, parallel with the spinal column. This systemic trunk (Figs. 108, 109) divides and subdivides into finer and finer arteries, like the branches from the trunk of a tree, excepting that these branches communicate with each other in a finer network till the ultimate ramifications, too minute to be seen by the naked eye, extend to every nook

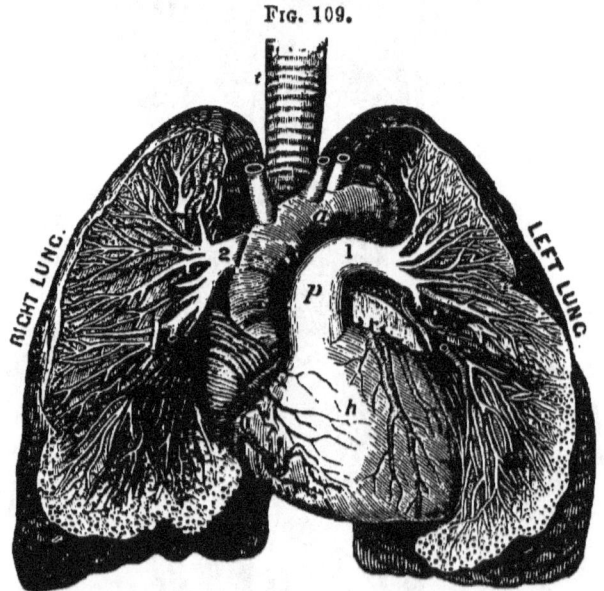

FIG. 109.

FIG. 109. THE PULMONARY ARTERY.—*t*, The trachea. *h*, The heart. *a*, The aorta. *p*, The pulmonary artery. 1, The branch of the pulmonary artery that divides in the left lung. 2, The branch that divides in the right lung.

and corner and atom of the body. These final branches are called *Capillaries*.

Observation.—The name given to the aorta in the chest is Thoracic It the abdomen it is named the Abdominal aorta. In the sacral part of the abdomen it finally separates into two divisions, called Iliac arteries. In the thigh, above the knee, its sub-division is named Femoral; below the knee, Anterior and Posterior Tibial arteries. From the Arch of the Aorta there are given off several large branches—the Carotid, which carries blood to the head; the Subclavian; its branches in the

State the names of the Aorta in different parts of the body.

arm are named Brachial; below the elbow, Radial and Ulnar arteries. (Figs. 108, 115.)

221. The PULMONARY ARTERY commences in front of the origin of the aorta. It ascends obliquely to the under surface of the arch of the aorta, where it divides into two branches, one of which passes to the right, the other to the left lung. These divide and subdivide in the structure of the lungs and terminate in the capillary vessels, which form a network around the air-cells and become continuous with the minute branches of the pulmonary veins. (Fig. 109.)

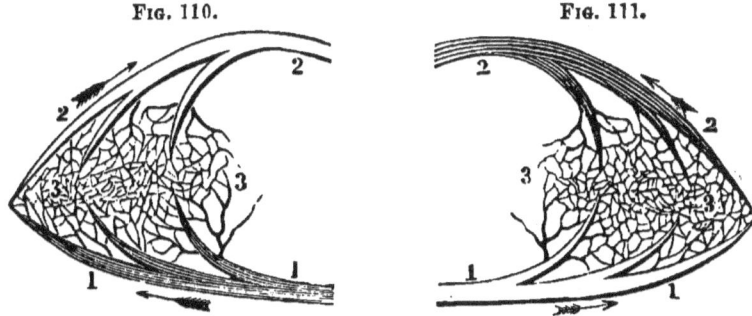

FIG. 110. AN IDEAL VIEW OF A PORTION OF THE PULMONIC CIRCULATION.—1, 1, A branch of the artery that carries the impure blood to the lungs. 3, 3, Capillary vessels. 2, 2, A vein through which red blood is returned to the left side of the heart.

FIG. 111. AN IDEAL VIEW OF A PORTION OF THE SYSTEMIC CIRCULATION.—1, 1, A branch of the aorta. This terminates in the capillaries (3, 3). 2, 2, A vein through which the impure blood is carried to the right side of the heart.

222. The CAPILLARIES serve to connect the termination of the arteries with the beginning of the *veins*, so that it is impossible to tell just where the artery ends and the vein begins. In these minute vessels the blood comes in intimate relation with the substance of the tissues, making them the most important part of the whole circulatory system. The operations of secretion and the conversion of the nutrient materials of the blood into bone, muscle, etc., are performed in these vessels. (Figs. 110, 111.)

223. The VEINS thus commencing with the capillaries unite into larger and larger veins, converging *toward* the

---

Describe the Pulmonary Artery. The Capillaries. Where do the veins commence. Give the course of the veins.

heart till the final union in two trunks, the *Ascending* and *Descending Venæ Cavæ*, that connect with the right auricle of the heart.

Observation.—The Ascending Vena Cava collects the blood from the lower extremities, pelvis and abdomen, and terminates in the right auricle of the heart. The Descending Vena Cava derives its branches from the head, neck, upper extremities and walls of the thorax. It terminates at the upper back part of the right auricle of the heart. The Aorta and Cavæ constitute the large vessels of the *Systemic* or *General Circulation.* (Fig. 108.)

224. The PULMONARY VEINS are four in number, two for each lung. They commence with the capillaries of the lungs and converge till a single trunk is formed for each lobe, or three trunks for the right lung and two for the left; but the trunk from the middle lobe of the right lung joins that from the upper lobe of the same side, and the four mouths discharge into the four angles of the left auricle. These Veins, with the Pulmonary Artery, establish the *Lesser* or *Pulmonic Circulation.*

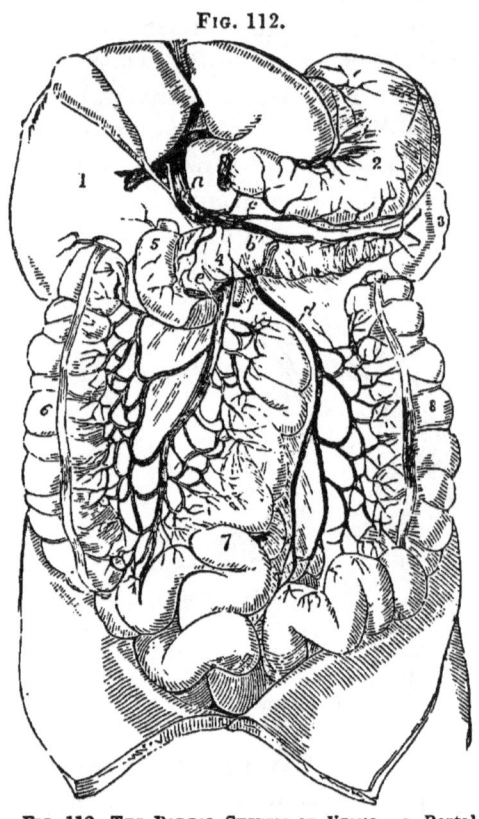

FIG. 112. THE PORTAL SYSTEM OF VEINS.—*a*, Portal vein. *b*, Splenic vein. *c*, Right gastro-epiploic vein. *d*, Inferior mesenteric vein. *e*, Superior mesenteric vein. *f*, Trunk of the superior mesenteric artery. 1, Liver. 2, Stomach. 3, Spleen. 4, Pancreas. 5, Duodenum. 6, Ascending colon; the transverse colon is removed. 7, Small intestine. 8, Descending colon.

What vessels constitute the Systemic Circulation? Describe the Pulmonary Veins. What constitutes the Pulmonic Circulation?

225. The PORTAL VEIN (so called because it enters the liver by a kind of gateway upon its under surface) is a short trunk about three inches in length, derived from the convergence of the veins of the stomach, spleen, pancreas and intestines; this passes into the liver, where it divides and subdivides, being distributed throughout the organ. This blood is returned from the liver to the general circulation by the hepatic veins. (Fig. 112.)

25. PHYSIOLOGY OF THE CIRCULATORY ORGANS.—*Necessity for Circulation—For the Double System of Circulation. Plan of Systemic Circulation—Of Pulmonic Circulation—Their Relation to Each Other.*

226. The Tissues are so constructed that their vitality depends upon their activity, and their activity upon the amount of oxygen and nutritive material supplied, the oxygen being essential to the chemical combinations, without which there could be no new deposit of tissue particles, and the nutritive matter being necessary to supply the waste produced by these chemical and vital activities; hence the necessity of a pneumatic (air) apparatus for providing a constant and sufficient supply of oxygen, and of a hydraulic (fluids in motion) apparatus for conveying the prepared nutriment to every atom of the body, and also to remove the waste, worn-out particles. The former need is met by the exquisite mechanism of the lungs, and the latter by the no less refined mechanism of the heart and blood-vessels. The two apparatuses are brought into use and harmonious co-working by the *double circulation* of the blood; hence the necessity of the double heart. (Figs. 107, 115.)

227. From the *left ventricle* the blood is forced into the aorta, to be diffused through the arteries to the capillaries of every part of the body; thence it is returned by the veins, through the venæ cavæ, to the *right auricle*, which delivers it to the right ventricle; this completes the *Systemic Circulation*. (Figs. 115, 116.)

Describe the Portal Vein. Why is Circulation necessary? Why a double heart? Give the Systemic Circulation.

228. From the *right ventricle* the blood is thrown into the pulmonary artery, and through its branches to the pulmonary capillaries, thence returned by the pulmonary veins, which coalesce into four trunks, and finally enters the *left auricle*, which immediately pours it into the left ventricle. This completes the *Pulmonic Circulation*, and the two constitute one complete circuit of the double circulation. (Figs. 113, 114.)

229. Both circulations are carried on at the same time—that is, the auricles contract and dilate simultaneously; the

FIG. 113.   FIG. 114.

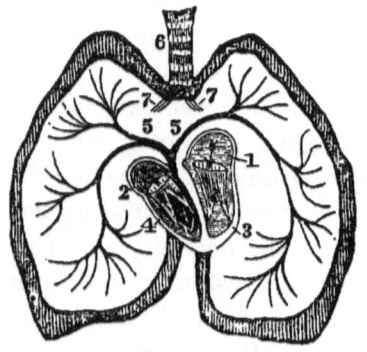 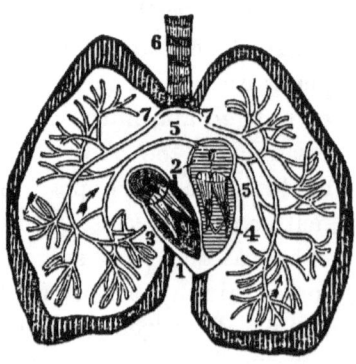

FIG. 113. A DIAGRAM.—1, Left auricle. 2, Right auricle. 3, Left ventricle. 4, Right ventricle. 5, 5, Pulmonary artery. 6, Trachea.

FIG. 114. A DIAGRAM.—1, Right auricle. 2, Left auricle. 3, Right ventricle. 4, Left ventricle. 5, 5, Right and left pulmonary veins. 6, Trachea.

same is true of the ventricles, whose action immediately follows that of the auricles. Hence, at the same instant, by the action of the ventricles, pure blood is thrown into the body and impure blood into the lungs; and at the same instant the right auricles receive impure blood from the body and the left auricle pure blood from the lungs.

§ **26.** HYGIENE OF THE CIRCULATORY ORGANS.—*Conditions favoring Free Circulation. Observation.*

230. *A natural and equal temperature should be preserved.* The blood-vessels are contracted by cold; thus a chill in

---

Give the Pulmonic Circulation. What is said of the contraction and dilatation of the auricles and ventricles? What is the effect of such action? Why should an equal temperature be preserved?

THE CIRCULATORY ORGANS. 123

any part of the body drives the blood to other parts. The chilled part is thus weakened, while the over-burdened parts suffer from congestion. If the surface is chilled, the blood

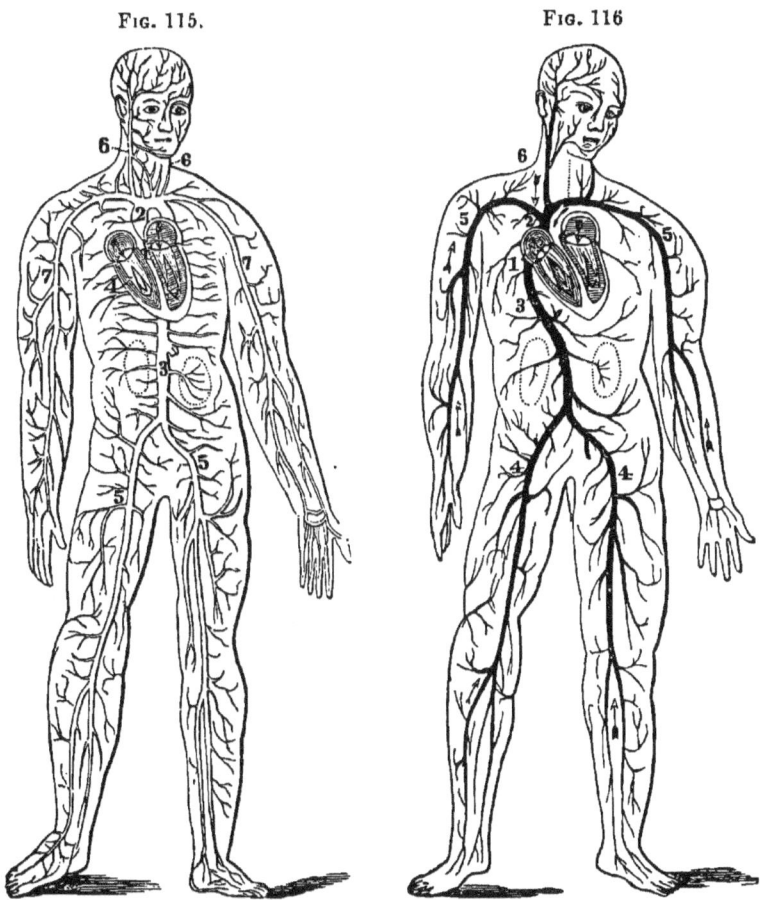

FIG. 115. A DIAGRAM.—1, Left ventricle of the heart. 2, 3, Aorta. 5, 5, Arteries that extend to the lower extremities. 6, 6, Arteries of the neck. 7, 7, Arteries of the arms.
FIG. 116. A DIAGRAM.—1. Right auricle of the heart. 2, 3, Large veins that open into the right auricle. 4, 4, Veins of the lower extremities. 5, 5, Veins of the arms. 6, Veins of the neck. The arrows show the direction that the blood flows.

is thrown upon the internal vital organs; hence the necessity of warm clothing and also frequent bathing, which favors the free action of the vessels of the skin.

231. *The clothing should be loosely worn.* Compression of any kind impedes free circulation. Pressure about the vital organs is especially injurious. Ligatures used to retain in place any article of apparel should be elastic. Tight dressing of the neck deprives the brain of its due amount of blood and retards the free return of venous blood from this organ—an item of particular importance to students, public speakers and persons predisposed to apoplexy or any brain disease.

232. *Exercise promotes the circulation of the blood.* By the action of the muscles the blood is propelled more rapidly through the blood-vessels, thus promoting a vigorous circulation in the extremities and skin. The best stimulants for a pale skin and cold extremities are a union of vigorous muscular exercise with agreeable mental action, and systematic bathing attended by thorough friction.

233. *The quality and quantity of the blood modify the action of the heart and blood-vessels.* If this fluid is abundant and pure, the circulatory vessels act with more energy than when it is deficient in quantity or defective in quality.

Observation.—If blood in large quantities is drawn from the veins, the heart will beat feebly and the pulse become weak. A similar effect is produced when the blood from any cause becomes vitiated.

§ **27.** COMPARATIVE ANATOMY (Angiology).—*Circulation of the Blood in Mammals—Birds—Reptiles—Amphibians—Fishes—The Annulosa—Insects—The Mollusca—The Radiata.*

234. The blood of *Mammals* is red, and the globules generally round. In the Camel they are elliptical. The hearts of Mammals have two auricles and two ventricles. The heart in quadrupeds lies on the middle line of the body, and not a little to the left of it, as in man. There is a marked peculiarity in the distribution of the arteries of quadrupeds. In the long necks of grazing animals there is found a large number of small arterial trunks, which are termed "Wonder Nets."

Why should the clothing be worn loosely? What is the influence of exercise on circulation? What is said of the quality and quantity of the blood? Observation. What is said of the blood and circulatory organs of Mammals?

Were these trunks few and large, as in man, the life of the animal would be endangered by the constant dependent position of the head.

235. The blood of *Birds* has the highest temperature of the vertebrate animals. It is richer in the blood cells (corpuscles) than in man, and these are elliptical. The heart of birds is highly muscular, and of large size in proportion to the bulk of the body. The aorta, at its commencement, divides into three large branches, of which the first two convey the blood to the head and neck, wings, and muscles of the chest; while the third, curving downward around the right bronchial tube, becomes the descending aorta. There are "Wonder Nets" in various parts of the body, especially in the arteries supplying the brain, eyes and legs.

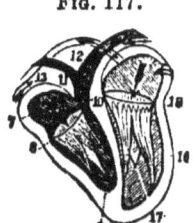

FIG. 117.

FIG. 117. DIAGRAM OF THE HEART OF THE MAMMAL.—7, Right auricle. 8, Right ventricle. 10, Pulmonic artery. 12, Pulmonic vein. 15, Left auricle 16, Left ventricle.

236. In *Reptiles* the blood is cold; that is, only slightly warmer than the temperature of the air or water in which they live, having fewer globules and lighter in color. The heart consists of two auricles and one ventricle. The arterial blood coming from the lungs is received into the left auricle, and the venous blood from all parts of the body into the right auricle; both are poured into the single ventricle, thus mixing the pure and impure blood, which will account for the sluggishness of these animals. A portion of this mixture returns by the aorta into the different organs it is intended to nourish, while another part proceeds to the lungs by vessels springing from the ventricle or the aorta. (Fig. 119.)

237. The *Amphibians* are cold-blooded animals. The blood-globules are larger than in mammals. Their circulation is incomplete, as in reptiles.

238. In *Fishes* the blood is cold, usually red, and the globules small. The heart has one auricle and one ventricle,

---

What is said of the blood and circulatory organs of Birds? Of Reptiles? Of Amphibians? Of Fishes?

126   ANATOMY, PHYSIOLOGY AND HYGIENE.

containing only impure blood; this blood is sent to the gills, which answer the purpose of lungs, and after being there exposed to the oxygen of air contained in the water and purified, it is distributed immediately to the different parts of the

Fig. 118.

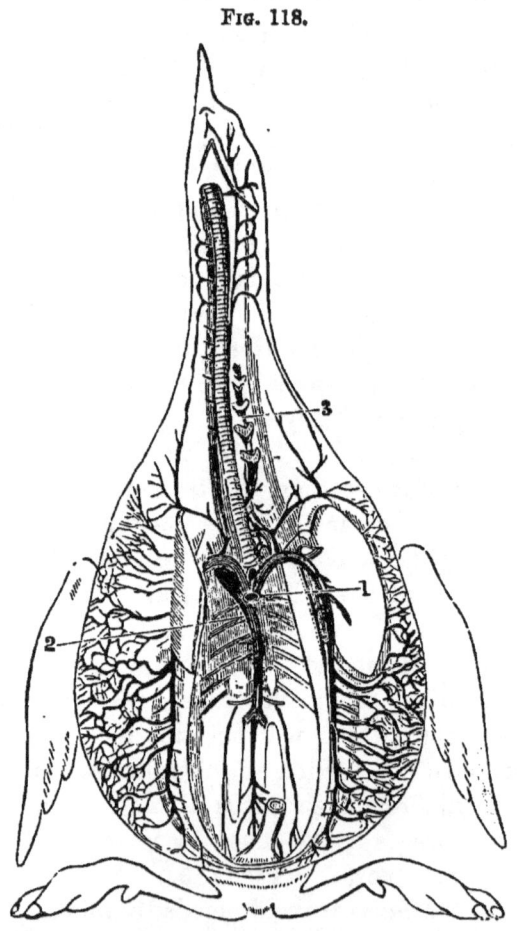

Fig. 118. ARTERIES OF THE TRUNK OF A BIRD.—1, The aorta. 2, The vena cava. 3, A cerebral artery. The small lines on each side represent the arteries and veins of the lungs.

body without the interposition of a heart. From the body the blood is returned to the auricle.

239. In the ANNULOSA the circulatory system is incomplete,

What is said of the blood and circulatory organs in the sub-kingdom Annulosa?

and occasionally altogether absent. In general, the blood is white or purple; in some species the circulating fluid is red, green, blue or a yellow-brown In the class Crustacea, of which the Crab and Lobster are examples, there is a single ventricle, which receives the blood from the gills and propels it to other parts of the body. The veins are everywhere replaced by irregular cavities called venous *sinuses*. (Fig. 121.)

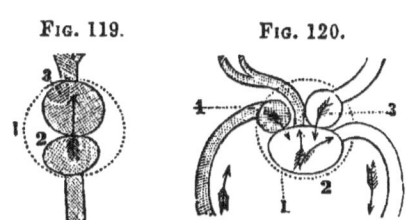

FIG. 119. DIAGRAM OF THE HEART OF THE REPTILE.—1, Pericardium. 2, Single ventricle. 3, Left auricle. 4, Right auricle. The arrows show the direction of the blood.

FIG. 120. DIAGRAM OF THE HEART OF THE FISH.—1, Pericardium. 2, The ventricle that receives the blood from the body. 3, The ventricle that sends blood to the gills.

Insects have neither arteries nor veins. The circulation, such as it is, is animated by the action of a vessel called dorsal, which is situated above the digestive tube. (Fig. 122.)

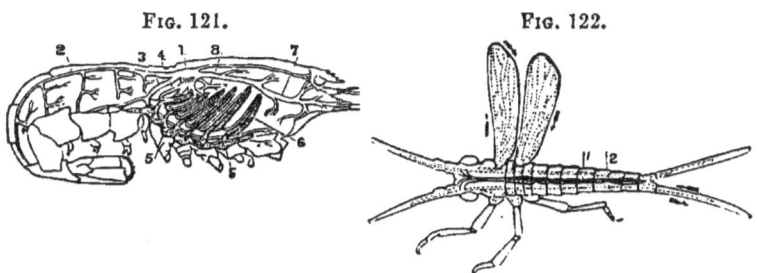

FIG. 121. THE HEART AND ARTERIES OF A LOBSTER.—1, The heart. 2, The abdominal artery. 5, 5, Venous sinuses. 6, The branchiæ from which the blood returns to the heart.

FIG. 122. DIAGRAM OF THE CIRCULATION OF AN INSECT.—1, Dorsal vessels in which the blood flows. 2, The lateral currents. The arrows show the outward and the inward current.

240. In the higher orders of the MOLLUSCA the circulation resembles that of fishes. The heart has usually a ventricle, whence spring the arteries and one or two auricles which

---

What is said of the blood and circulatory organs in Lobsters and Crabs? In Insects? In the Mollusca?

128 ANATOMY, PHYSIOLOGY AND HYGIENE.

FIG. 123. THE HEART AND ARTERIES OF A SNAIL.—2, The Stomach. 3, Intestines. 5, Heart. 6, Aorta. 7, Pulmonary artery

carry the arterial blood from the respiratory apparatus, which this liquid reaches by venous tubes more or less complete. The blood varies in color, and is generally destitute of globules.

241. In the RADIATA those highest in organization have a heart and vessels corresponding in some degree to the Mollusca. In the Infusoria (class Protozoa) the *contractile vesicle* appears to represent a rudimentary circulatory system.

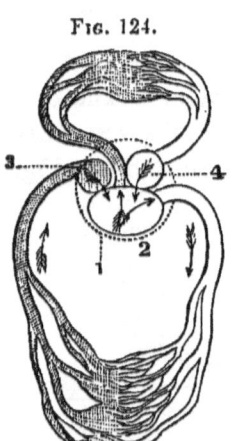
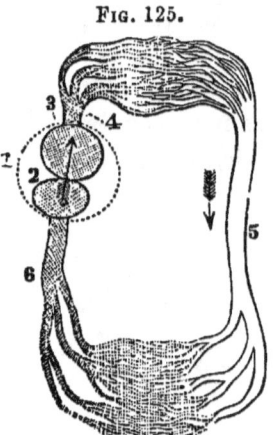

FIG. 124. AN IDEAL PLAN OF THE CIRCULATION OF A FROG.—1, The pericardium. 2, The single ventricle. 3, The right auricle. 4, The left auricle. The arrows indicate the direction of the blood.

FIG. 125. AN IDEAL PLAN OF THE CIRCULATION OF A FISH.—1, The pericardium. 2, The single auricle. 3, The single ventricle. 4, The vessel that conveys the blood from the ventricle to the gills. 5, The vessel that conveys the blood from the gills to the body of the fish. 6, The vessel that conveys the blood from the body to the heart. The arrows show the direction of the blood.

What is said of the blood and circulatory organs in the Radiata? In the Infusoria?

THE CIRCULATORY ORGANS.

Fig. 126.

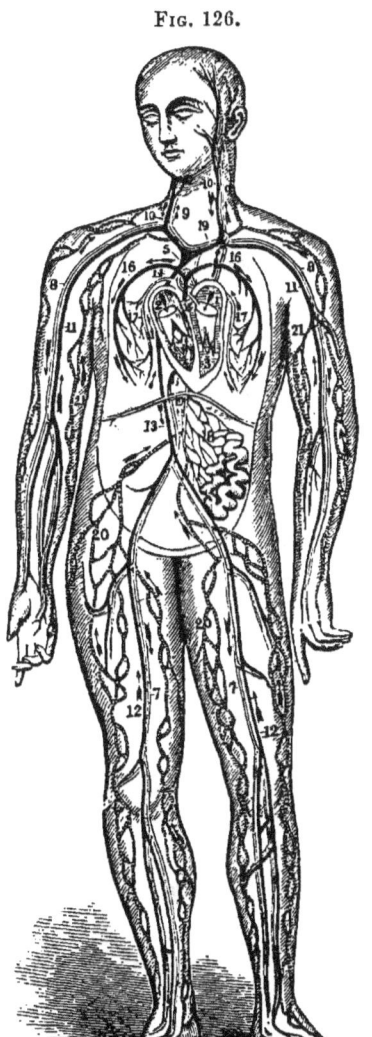

Fig. 126.—1, The left auricle. 2, The right auricle. 3, The left ventricle. 4, The right ventricle. 5, 6, The aorta. 7, 7, The femoral (leg) arteries. 8, 8, The brachial (arm) arteries. 9, 9, The carotid arteries. 10, 10, The jugular veins. 11, 11, The brachial veins. 12, The femoral veins. 13, 14, The vena cavas. 15, The pulmonary artery. 16, 16, The right and left pulmonary arteries. 17, 17, The pulmonary veins. 18, The lacteals. 19, 19, The thoracic duct. 20, 20, The absorbents of the lower extremities. 21, 21, The absorbents of the upper extremities. 22, The small intestine.

SYNTHETIC TOPICAL REVIEW
OF THE CIRCULATORY SYSTEM.
Classes, Sub-kingdoms, Divisions, Anatomy, Physiology, Hygiene.

State the Anatomy, the Physiology, the Hygiene of the Circulatory System. Human and Comparative.

F *

CHAPTER IX.

ASSIMILATION.

§ 28. ASSIMILATION.—*Changes included under Secondary Assimilation. Secretion. Excretion. The Kidneys. Animal Heat.*

242. As before stated, the processes by which food is changed into chyle and then into blood may be included under *Primary Assimilation*. The changes which convert portions of the blood into solid tissue may be termed *Secondary Assimilation*.

243. SECONDARY ASSIMILATION (Nutrition of the Organs and Tissues) consists of the following stages: *First*, A nutritive fluid (*plasma*) exudes from the blood through the coats of the capillaries, filling the finest spaces of the tissues between the capillary networks, and bathing all the elementary parts of these tissues. The nature of this fluid is the same in all parts of the system.

Second, The nutritive process consists in the exercise of a certain selective act by the elementary parts of tissues and organs, enabling them to appropriate to themselves such portions of the nutritive fluid as are suitable, either with or without further change, to renew, particle by particle, their worn-out substance.

Third, The result of the act of assimilation is to leave a residual fluid in the interspaces of the tissue-elements outside the capillary vessels. The nature of this fluid must differ in the different tissues, inasmuch as different tissues make different appropriations.

Fourth, The final residuum, or that which is not taken up by the tissues or lymphatics, is probably taken up by the venous capillaries.

Distinguish between Primary and Secondary Assimilation. State the first stage in the nutrition of the organs and tissues. What is the second? The third? The fourth?

244. *Assimilation* is the formation of the tissues from the elements of the food. *Secretion* is the separation from the blood of materials in a more or less fluid condition through a *Gland* or *Membrane*. After assimilation or secretion the products are discharged from the ducts of the glands or the surfaces of the membranes, and are used for certain purposes in the animal economy, or excreted from the body.

245. The SECRETING GLANDS are the liver, the pancreas, the salivary and the lachrymal glands; the true mucous glands of the nose, mouth, fauces, pharynx, œsophagus, duodenum, and those of the skin, some of the glands of the stomach and intestines and the sebaceous glands.

246. The SECRETING MEMBRANES are the mucous, which lines all those passages that communicate with the air; the serous, which covers the organs of the body not exposed to the air, and also lines the cavity in which those organs are contained; and the synovial membranes, which compose a part of a joint.

247. With the final residuum are mingled the worn-out particles of waste from the tissues, which also enter the venous blood to be thrown off in the form of *excretions*.

248. EXCRETION is *effected by glands only*, and the exhausted, useless particles are expelled from the blood and thrown out of the system. The excretory glands are the kidneys, the sweat glands of the skin, to a certain extent the liver, the sebaceous glands of the skin, and lastly, the lungs, which throw off carbonic acid from the blood.

249. The KIDNEYS lie one on each side of the spinal column. Their shape is that of a bean, and their color a brownish red. The substance of the kidney is mainly composed of secretory tubes. In these small tubes or *tubules* the urine is secreted. The duct to the bladder is called *Ureter*. The kidneys receive a very large supply of blood, and they are the only glands that throw off certain injurious substances from the blood.

What is Assimilation? What is Secretion? Name the Secreting Glands and Membranes. What is mixed with the venous blood? How is Excretion effected? Name the Excretory organs. Describe the kidneys. What is their office?

132   ANATOMY, PHYSIOLOGY AND HYGIENE.

Observation.—The retention of the secretion of the kidneys should never be allowed by the young or the old, the healthy or the diseased, as suppression of the secretion of these glands immediately affects the whole system, especially the nervous centres. Both the quantity and color of this secretion indicate the condition or health of the body.

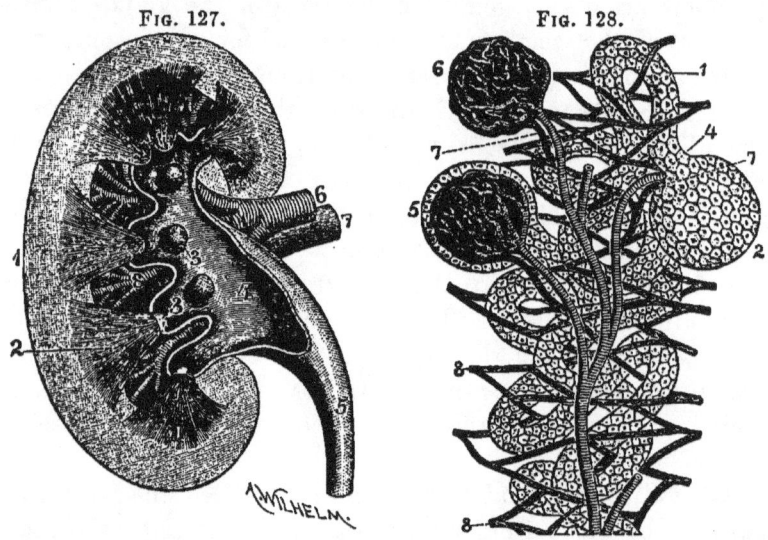

FIG. 127. LONGITUDINAL SECTION OF A KIDNEY.—1, 2, 3, Parts of the Kidneys. 4, Pelvis. 5, Ureter. 6, Renal artery. 7, Renal vein. 8, Branches of the latter vessels in the kidney.

FIG. 128. DIAGRAM OF THE STRUCTURE OF THE KIDNEYS.—1, Tubules or minute tubes. 2, Enlargement of a tubule at its extremity. 3, Branches of the real artery ending in vessels which enter the enlargements as seen at 4, 5. 6, Knot of blood-vessels freed from its investment. 7, Veins emerging from the knots. 8, Plexus formed by the latter veins among the tubules, from which plexus originate the branches of the renal vein.

250. The nutritive changes going on within the animal body is the source of its heat. In man and most mammals the interior of the body is about 100°, whether in the torrid zone or within the polar circle. The external parts may be frost-bitten, still the internal temperature or *Animal Heat* is nearly invariable.

Observation.—There is a difference in the heat-producing powers of different animals. In the young of all animals the temperature is a few degrees higher than in the adult, and breathing is correspondingly more frequent, but the sensibility to cold is much greater.

What is the source of the heat of the body? What is the temperature of the body? Why do young animals need more protection or clothing than those of mature growth?

DIVISION III.

SENSORIAL APPARATUS.

251. In the two preceding Divisions the tissues and organs directly involved in the movements of the body, and those most intimately connected with the preparation and assimilation of nutrient material, have been briefly described. In the present Division we consider the organs through which is manifested the subtle power that controls these motions and processes, establishes telegraphic communication between the several parts of the body and brings it into important relations with the external world. These, taken collectively, we name the SENSORIAL APPARATUS. The organs of this apparatus are the *Brain, Spinal Cord* and *Nerves*.

CHAPTER X.

NERVOUS SYSTEM.

§ **29.** *Two forms of Nervous Tissue. Classification of the Ganglions. The Nerves. The Spinal Cord. The Medulla Oblongata. The Cerebrum. The Cerebellum. Membranes of the Brain—The Nerves. Classification of Cerebro-Spinal Nerves—Of Cranial Nerves—Of Spinal Nerves. Sympathetic System.*

252. NERVOUS TISSUE presents two characters—one, cell-like and gray in color, arranged in masses called *Centres* or *Ganglions*, being the originating, active centres of nerve-force; the other, fibrous and white, arranged in threads or small cords called *Nerves*, which are simple conductors of nerve-force.

Name the organs of the Sensorial Apparatus. What two characters does Nervous Tissue present? Give the arrangement and names of each.

253. The GANGLIONS (Ganglia), a collection of nerve-cells, may be arranged in two great systems, the *Cerebro-Spinal* (brain and spinal cord) and the *Sympathetic* (ganglionic) with their nerves. (Fig. 129.)

254. The CEREBRO-SPINAL system commences with that portion of nervous matter which is enclosed in a long canal within the spinal column, extending from the second vertebra of the loins to the base of the skull, and known as the *Spinal Cord.*

255. The SPINAL CORD (marrow) is nearly round and double, the two halves connected by a narrow bridge (commissure) of the same substance as the cord, which is soft and white externally, but grayish within. The amount of white substance exceeds the gray.

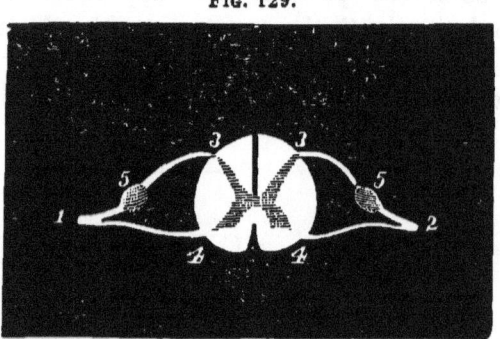

FIG. 129. TRANSVERSE SECTION OF THE SPINAL CORD.—1, 2 Spinal nerves of right and left sides, showing their two roots. 3, 3, Origin of posterior root. 4, 4, Origin of anterior root, with the commissure or bridge between the two halves. 5, Ganglion of posterior root.

A vertical fissure partly separates the cord into two halves, and each lateral half consists of two separate bundles of fibres, which are called the *Anterior* and *Posterior columns.* As the cord enters the cavity of the skull it becomes greatly enlarged, and receives the name of *Medul'la Oblonga'ta.* (Figs. 130, 135.)

256. The MEDULLA OBLONGATA preserves the connection between the spinal cord and the brain. Its columns are continuous with those of the cord, but contain a larger quantity of gray substance. Motory impressions are carried through

---

What are Ganglions? How arranged? Where does the Cerebro-Spinal system commence? Describe the Spinal Cord. Give its form after it enters the skull. Describe the Medulla Oblongata.

its anterior, and sensitive impression its posterior columns. No other part of the body is so immediately important to the maintenance of life as the Medulla Oblongata.

257. The BRAIN is the large mass of nervous tissue within the skull-bones. It consists of two parts, the *Cer'ebrum* (large brain), the *Cerebel'lum* (small brain).

FIG. 130.

FIG. 130. A SECTION OF THE BRAIN AND SPINAL COLUMN. —1, The cerebrum (large brain). 2, The cerebellum (small brain). 3, The Medulla Oblongata. 4, 4, The spinal cord in its canal.

258. The CEREBRUM, or larger portion of the brain, is composed of a whitish substance, with an irregular border of gray matter around its edges. It is divided into two hemispheres, and each hemisphere has three lobes, and each lobe has eminences and depressions which produce those winding inequalities called *convolutions*. (Fig. 130.)

259. The CEREBELLUM is also composed of white and gray matter, but the latter constitutes the largest portion. The white matter is so arranged that when cut vertically the appearance of the trunk and branches of a tree (*ar'bor vi'tæ*) is presented. (Fig. 132.)

260. The brain is surrounded by three membranes. The external membrane (*Dura Mater*) is thick and firm; the middle (*Arachnoid*) is thin, and looks somewhat like a spider's web; the inner (*Pia Mater*) consists of a network of blood-vessels. These membranes are prolonged so as to form a sheath to the spinal cord.

261. We find distributed throughout the body small white threads, or *nerve-filaments*, which interlace and loop them-

---

Describe the Brain. Describe the Cerebrum. What are irregularities of its surface called? Describe the Cerebellum. Speak of the membranes of the Brain. Describe the Nerve-filaments.

selves with the various tissues, reaching every fibre of a muscle, the papillæ of the skin and all the glandular organs. From these points they approach each other, uniting into little bundles or fibres, and then into larger bundles, till they are of sufficient size to be seen by the naked eye, when they constitute a *Nerve*.

The filaments do not blend with each other, but remain distinct from their origin to their termination. Like the fibres of a muscle, they are bound together and protected by a covering of tissue called its *Neurilem'ma* or sheath, which also contains the blood-vessels for the nutrition of the nerve.

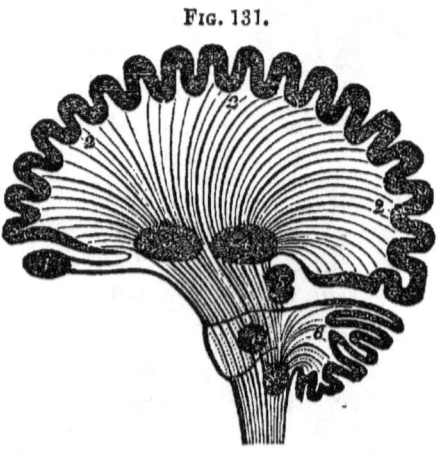

Fig. 131.

FIG. 131. DIAGRAM OF HUMAN BRAIN IN VERTICAL SECTION, showing the situation of the different ganglia and the course of the fibres.—1, Olfactory ganglion. 2, Hemisphere. 3, Corpus striatum. 4, Optic thalamus. 5, Tubercula quadrigemina. 6, Cerebellum. 7, Ganglion of tuber annulare. 8, Ganglion of medulla oblongata.

Observation.—The nerves have no power of originating impressions. The *Sensory* nerves do not originate in the brain; they are stimulated by *external agents*, and the impression is carried from the extremities to the nervous centres. The *Motory* nerves are stimulated by the *Will* or some other force generated in the nervous centres and conveyed *from* them to the distant parts of the extremities, chiefly the muscles.

262. The CEREBRO-SPINAL NERVES are called *Cranial* when they pass directly from the brain through the openings in the skull; and *Spinal* when they pass through the openings of the spinal column.

263. The CRANIAL NERVES are arranged in twelve pairs,

When are they called Nerves? What is said of the blending of the nerve-filaments? Give the name of the nerve-sheath. What does it contain? Observation. Give the arrangement of the Cerebro-Spinal Nerves. How are the Cranial Nerves arranged?

and these in three groups according to their functions, as *Sensory*, *Motory* and *Mixed*.

Name.	Destination.
1st pair..Olfactory..................	Mucous membrane of nasal passages.
2d " ..Optic.....................	Retina of eye.
8th " ..Auditory...................	Internal ear.

3d pair..Oculo-Motor................	The muscles of eye, excepting external rectus and trochlear.
4th " ..Patheticus................	Trochlear muscle of eye.
6th " ..Abducent.................	External rectus of eye, turning eye upward.
7th " ..Portio Dura..............	Different muscles of face, giving expression.
11th " ..Spinal Accessory...........	Muscles of neck.
12th " ..Hypo-glossal..............	Muscles of tongue.

5th pair..Trifacial..	Motor branches to muscles used in mastication. Sensory " to the teeth, tongue and different parts of the face. (Fig. 135.)
9th " ..Glosso-pharyngeal..........	Mucous membrane of the tongue and throat.
10th " ..Pneumogastric.............	Motor branches to the pharynx, larynx, trachea, lungs, heart, œsophagus, stomach and intestines. Sensory to ditto.

264. The SPINAL NERVES are arranged in thirty-one pairs. Each of these nerves arises by two roots—an anterior or *Motor* root, springing from the anterior columns of the spinal cord, and a posterior or *Sensitive* root, from the posterior columns of the spinal cord. Both roots unite into one trunk, forming the spinal nerve, which passes out of the spinal column through its openings. The Spinal Nerves are divided into—

Cervical (Neck)...	8	pairs.
Dorsal (Back)..	12	"
Lumbar (Loins)..	5	"
Sacral (Pelvis)..	6	"

Observation.—A singular feature of the Spinal Nerves in their connection with the brain through the spinal cord is the crossing (decussation) of the Motor and Sensitive fibres of the spinal cord. Hence, the nerves of the right side of the body are connected with the left side of the brain, and those of the left side of the body with the right side of the brain.

265. The SYMPATHETIC SYSTEM is double, consisting of two chains of ganglions, one on each side of the spinal column, running through the deep parts of the neck into the chest and abdomen. These ganglions are composed of nerve-cells, and communicate with each other, with the spinal cord and with the internal organs—as the heart, lungs, stomach,

FIG. 132.

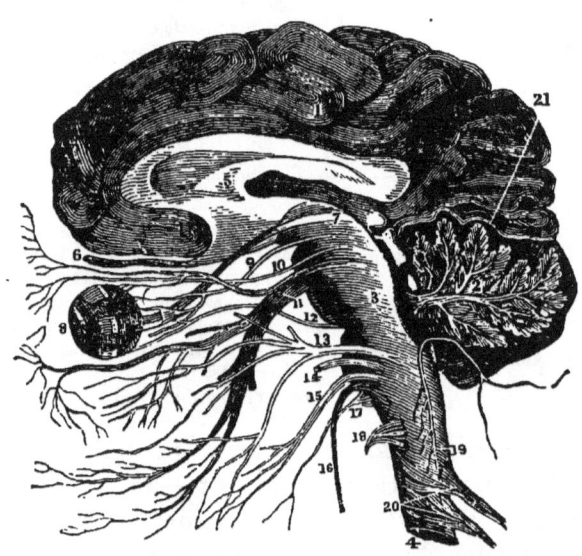

FIG. 132. A VERTICAL SECTION OF THE CEREBRUM, CEREBELLUM AND MEDULLA OBLONGATA, showing the relation of the cranial nerves at their origin.—1, The cerebrum. 2, The cerebellum with its arbor vitæ represented. 3, The medulla oblongata. 4, The spinal cord. 5, The corpus callosum. 6, The first pair of nerves. 7, The second pair. 8, The eye. 9, The third pair of nerves. 10, The fourth pair. 11, The fifth pair. 12, The sixth pair. 13, The seventh pair. 14, The eighth pair. 15, The ninth pair. 16, The tenth pair. 17, The eleventh pair. 18, The twelfth pair. 20, Spinal nerves. 21, The tentorium.

liver, pancreas, intestines and kidneys. Thus, the various parts of the body are associated with each other by a nervous apparatus which is only indirectly connected with the brain and spinal cord. (Fig. 135.)

A peculiarity of the Sympathetic Nerves is that they *follow the distribution of the blood-vessels.* Starting from the heart,

Describe the Sympathetic System. What is a peculiarity of the Sympathetic Nerves?

NERVOUS SYSTEM.

they envelop the large vessels with a close network called a *plexus*.*

Observation.—In all parts of the body these nerves accompany the arteries which supply the different organs, and form networks around them which take the names of the organs—as the hepatic plexus, splenic plexus, mesenteric plexus, etc.

§ **30.** PHYSIOLOGY OF THE NERVOUS SYSTEM.—*Classification of Nerve-centres—Ideational—Sensational—Reflex—Spinal—Organic Centres.*

266. The NERVOUS SYSTEM, though made up of several parts—as the Brain, Spinal Cord and Sympathetic Ganglion—yet each within certain limits performs distinct functions. To the hemispheres of the CEREBRUM, particularly the gray matter of its irregular border, may be ascribed the mental faculties—as Memory, Judgment and Reason. These ganglions or nerve-centres we call *Primary* or *Ideational.*

267. To the gray matter of the CEREBELLUM and

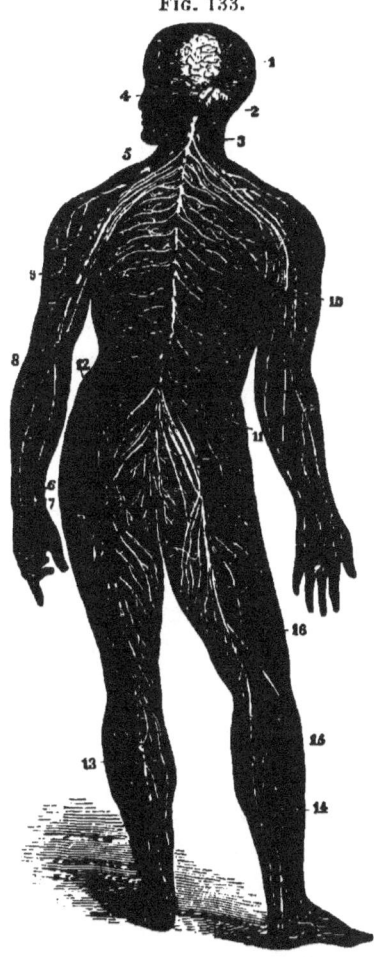

FIG. 133. A BACK VIEW OF THE BRAIN AND SPINAL CORD.—1, The Cerebrum. 2, The cerebellum. 3, The spinal cord. 4, Nerves of the face. 5, The brachial plexus of nerves. 6, 7, 8, 9, Nerves of the arm. 10 Nerves that pass under the ribs. 11, The lumbar plexus of nerves. 12, The sacral plexus of nerves. 13, 14, 15, 16, Nerves of the lower limbs.

What is a plexus? What is said of the functions of the different parts of the Nervous System? What is ascribed to the Cerebrum? What name is given to this centre?

* Lat., *plecto,* to intermingle.

other ganglions at the base of the brain is ascribed the seat of the higher and *special* senses, as sight and hearing. These ganglions are called the *Sensorial Centres.*

268. The SPINAL CORD is both a *conductor* of nervous impressions and a nervous centre of *reflex action.* Let an external agent touch any part of the surface of the body, the nervous stimulus is first conveyed *inward* by the fibres of the sensory nerves; it arrives at the spinal cord by the posterior or back roots, and reaches the gray matter in its central parts, which receives the nervous impression and instantly converts it into a motor impulse; it is then conveyed or *reflected outward* along the motor fibres of the anterior or front roots to the muscles. Such action is called the *Reflex Action* of the Nervous System. By this means a communication is established between the different organs. This communication is never direct, out from one organ inward to the nervous centre, then outward to another organ; so are the different functions associated and exercised for the common good of the whole. These may be called the *Reflex, Motory* or *Spinal Centres.*

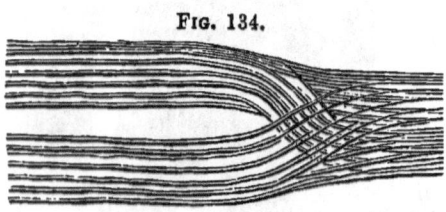

FIG. 134.

FIG. 134. NERVE-FILAMENTS, decussing with their sheath.

269. The SYMPATHETIC or ORGANIC CENTRES are not well understood, but the distribution of their nerves would indicate that they exercise a controlling influence over the involuntary functions of digestion, absorption, circulation and assimilation. These organic centres, being connected with the various organs by sensitive and motor nerves, are capable of an *independent reflex action.* They are also connected with the cerebro-spinal system, and are more or less assisted by and subordinate to it. In health the brain takes no cogni-

---

What is ascribed to the Cerebellum? What name is given to this centre? Name the functions of the Spinal Cord. What is Reflex Action? Give the name of the centres. Speak of the Sympathetic or Organic Centres. What is said of their connection?

zance of their action; when diseased, however, the centres report to the highest authority by means of cramps and other severely acute pains.

Observation 1.—Each nerve-centre acts independently within certain limits, but beyond these limits it is subordinate to the next higher; thus, the Organic Centres are subordinate to the Reflex or Spinal Centres, the Reflex to the Sensorial, and all to the Ideational or Supreme Centres. In each centre the individual cells probably differ in rank, some having a higher dignity, some a lower, but each its special appointment, its assigned duty.

2.—The mind is closely *united*, and yet it is distinct from the body through which it acts (dependent for its manifestations, but independent in essence). So intimate is the union that the body exercises a powerful influence in leading us upward into a true and higher life, or downward into a low and sensual existence. What this influence shall be depends somewhat upon *inherited organization*, but more upon *education*. Accepting the theory already advanced as at least illustrative, we see that if the thoughts, feelings and desires are pure and true and good, their impressions remaining in the nerve-cells are of the same character, and tend to give a right direction to the future activities of these cells. If the thoughts, feelings and wishes are evil in nature, the

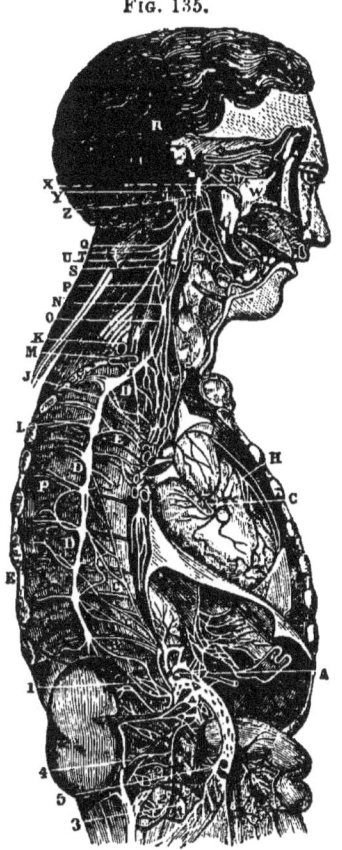

FIG. 135.

FIG. 135. THE SYMPATHETIC GANGLIONS AND THEIR CONNECTION WITH OTHER NERVES.—A, A, A, The semilunar ganglion and solar plexus. D, D, D, The thoracic (chest) ganglions. E, E, The external and internal branches of the thoracic ganglions. G, H, The right and left coronary (heart) plexus. I, N, Q, The inferior, middle and superior cervical (neck) ganglions. 1, The renal plexus of nerves. 2, The lumbar (loin) ganglions. 3, Their internal branches. 4, Their external branches. 5, The aortic plexus of nerves.

What is the relation between the sympathetic nerves and the brain? Observation 1. Speak of the relation of the mind to the body.

impressions will also be evil, inclining to evil activities in the future. When we resist a temptation to wrong action, then we not only avoid the particular evil, but lay up that which will render the next resistance easier and more natural. If we yield to the temptation, we are not only guilty of the particular wrong, but lay up that which will make resistance more difficult or yielding more easy and natural for the future. When a man sets his *heart to do right*, all his physical being struggles to give him aid; and when he sets his *heart to do wrong*, its energies are expended in dragging him downward.

3.—The *visible impress* which the workings of the mind leave upon the body is worthy our notice. The character of the man is declared by the lines of his muscles, which tell no lies. Especially is this true of the muscles of his face. Let him narrow his soul by penuriousness, become the victim of rasping jealousy, wear the nettles of envy against his heart, or be the slave of defiling lust, and in spite of any natural comeliness or studied concealment, his true character will be proclaimed to all who have learned aught of the language of the muscles. "Be sure your sin will find you out," says He who has made the fleshly lineaments to reveal the most hidden vice. The more secret the viciousness, the deeper is the impress. But if the spirit of evil thus leaves the traces of its blackened pen upon the face, the spirit of goodness writes thereon in no less legible characters of light. Purity of heart, nobleness of purpose, restfulness of soul, soften, irradiate, *spiritualize* the outer man, giving a higher beauty than that of form or complexion, even to him who is wrinkled by years, bowed by infirmity and scarred by the battles of life.

§ **31.** HYGIENE OF THE NERVOUS SYSTEM.—*Two Classes of Agencies affecting the Health of the Nervous System. Natural Heritage. Importance of the Physical Agency—Air—Diet—Exercise and Sleep. The effect of Mental Impressions on the Body. Mental Exercise. Recreation and Amusement. Harmonious Development of the Different Mental Powers.*

270. In considering the *Hygiene of the nervous system* it is necessary to have reference both to physical and mental agencies. The highest health and vigor of the nervous system doubtless require—1st, A sound nervous organization by inheritance; 2d, A nutrition equal to the demands of repair and growth; 3d, The harmonious action of the various mental powers.

Speak of the visible impress of the mind upon the muscles of the face. What agencies affect the health of the nervous system? Name the requirements of its health and vigor.

271. Every-day observation shows that children inherit not only the features, but the physical, mental and moral constitution, of their parents. Even those utterly ignorant of the laws of transmission are wont to estimate the child according to its family; favorably, if of a "good family" or "good blood;" unfavorably, if of a "bad family" or "bad blood."

Every formation of body, internal and external, all intellectual endowments and aptitudes, and all moral qualities, are or may be transmissible from parent to child. If one generation is missed, the qualities may appear in the next generation. It is important to notice that not only the *natural constitution* of the parents may be inherited, but their *acquired habits* of life, whether virtuous or vicious, but especially is this true of vice. Even when the identical vice does not appear, there is a morbid organization and a tendency to some vice akin to it. Not only is the evil tendency transmitted, but what was the simple practice, the *voluntarily* adopted and cherished vice of the parent, becomes the passion, the overpowering impulse, of the child.

Illustration 1.—M. Morel sketches the history of four generations as follows: "*First Generation.*—The father was an habitual drunkard, and was killed in a public-house brawl. *Second Generation.*—The son inherited his father's habits, which gave rise to attacks of mania, terminating in paralysis and death. *Third Generation.*—The grandson was strictly sober, but full of hypochondriacal and imaginary fears of persecutions, etc., and had homicidal tendencies. *Fourth Generation.*—The fourth in descent had very limited intelligence, and had an attack of madness when sixteen years old, terminating in stupidity nearly amounting to idiocy; with him the race probably becomes extinct."

2.—Says a learned physician, after long and close observation of the evil effects of *tobacco:* "If the evil ended with the individual who, by the indulgence of a pernicious custom, injures his own health and impairs his faculties of mind and body, he might be left to his enjoyment, his *fool's paradise*, unmolested. This, however, is not the case. In no instance is the sin of the father more strikingly visited upon the children than in the sin of tobacco-smoking. The enervation, the hysteria, the insanity, the dwarfish deformities, the consumption, the suffering

---

What, in addition to the features of parents, do children inherit? May acquired habits be transmitted? Illustrations.

lives and early deaths of the children of inveterate smokers bear ample testimony to the feebleness and unsoundness of the constitution transmitted by this pernicious habit."

3.—Should we trace the effects of the whole list of vices, it would be with equally sad results; even of the great love of money-getting, the celebrated Dr. Maudsley says: "I cannot but think, after what I have seen, that the extreme passion for getting rich, absorbing the whole energies of a life, does predispose to mental degeneration in the offspring, either to moral defect, or to moral and intellectual deficiency, or to outbreaks of insanity."

272. The relation of the nervous centres to the blood is the same in kind as that between other parts of the body and their blood-supply. Great waste is produced by nervous action; hence, the centres are very largely supplied with blood-vessels, especially the Ideational centres (hemispheres of the brain).

273. *The nervous system may be impaired by impure air.* Everybody knows that bad air injures the lungs, but few realize that, on the whole, it injures the brain still more. As the nerve-tissue is the most delicate part of the body, it soonest feels the evil effects of imperfectly purified blood.

274. *The nervous system may be impaired by improper diet.* We are wont to believe that improper diet may affect the digestive organs, but seldom consider the *mental* and *moral* effects of such diet. Improper food poisons the blood, and thus the nerve-centres are not only cheated of their nutriment, but are poisoned; hence, the ideas become confused, and the will weakened. The whole man is crippled, physically, mentally and morally. It is an indisputable fact that *bad bread,* for instance, may thus have a very *immoral* influence. Those much engaged in mental labor suffer most from bad diet. No teacher can teach well, no lawyer can plead well, no physician can practice well, no minister can think or preach well, who habitually takes improper food.

Observation 1.—If such be the effect of improper food, what shall we say of such poisons as alcohol, opium, haschish, tobacco, etc., which act

---

State the second requirement of health and vigor. Speak of the evil of breathing impure air What are the results of improper diet? Observations.

NERVOUS SYSTEM. 145

so directly and powerfully upon the nervous system? The same poison does not equally affect all the nerve-centres; thus, *strychnine* acts upon the spinal centres, but not the cerebral; *haschish*, upon the sensory centres, giving rise to hallucinations; *alcohol*, upon the cerebral centres particularly. The alcoholic poison first produces an increased activity of the muscles, then alternate exaltation and depression, both physical and mental; finally, stupor, relaxation of the muscles and deep sleep. These symptoms are transitory; but let the poisoning process be continued, and true delirium, so well known as "delirium tremens," follows, and at length what is known as "chronic alcoholism;" and, while intoxication lasts a few hours, and delirium tremens a few days or weeks, chronic alcoholism spreads its baneful influence over years, unless death prevents the full development of the tragedy. The victim of alcoholic poison is equally enfeebled in body and mind. The nervous system becomes exhausted, the moral sentiments perverted, the will-power broken, and he seems powerless to cease from the fatal habit which has produced the change.

2.—With the *opium-eater* the diseases of the nervous system declare themselves even more rapidly than with the drunkard. Says M. Morel: "Given the period at which a person begins to smoke opium, and it is easy to predicate the time of his death: his days are numbered."

3.—*Tobacco* is one of the most virulent poisons. It soothes the nerves temporarlly, only to leave them more enfeebled and irritable.

4.—Even excessive use of *tea* and *coffee* may prove disastrous to the health of the nervous system.

275. *The nervous system may be impaired by want of physical exercise.* Among other agencies that affect the nervous system, none exert a wider influence than bodily exercise. It seems to be required to complete the change which the blood undergoes while passing through the lungs and skin, without which the waste of nerve-element could not be repaired. In persons who are merely sedentary, having little occasion for active thought, this want of exercise is sufficiently mischievous; but when there is great mental activity, the mischief is vastly increased.

Observation.—Thousands of ministers, lawyers, those who sit in the bank and counting-room, shorten their days because of this neglect: especially is this the case in America. The English nobility, notwithstanding their many indulgences, are a long-lived race, and this is doubtless owing to their spending so much time in open-air exercise.

276. *The nervous system and mental activities may be enfeebled by an unhealthy skin.* If its healthy state is impaired by want of cleanliness, by deficient clothing or by a diseased action of the cells of the skin (the nucleated epithelium), through an intimate sympathy in like tissues, the cells of the nerve-tissue may be seriously affected.

277. *The nervous system may become impaired by taking too little sleep.* "Sleep knits up the raveled structure" of nervous element, for during sleep organic assimilation is restoring what has been expended in functional energy. A renewal of nervous energy as often as once a day is an institution of Nature.

Observation 1.—Among the wise arrangements of the Creator, none harmonize with the wants of the system more perfectly than the alternation of day and night. The amount of sleep necessary depends upon the age, health, natural temperament and occupation of the individual. The more rapid the exhaustion of nervous energy from any cause, the more sleep will be required. The young and the aged need more sleep than the person of middle life, the sick more than the well, those engaged in mental pursuits more than those wearied by manual labor, persons of great sensibility more than the sluggish natures whose normal condition is more nearly allied to sleep, woman more than man. We may say in general that the time should not be less than from six to eight hours, and most persons require a longer period. The time, however, must be proportioned to the need.

2.—Among the wealthy classes the customs of the times are quite at variance with those habits of sleep which are essential to mental vigor. Where amusements are pursued till late hours night after night, the nervous system greatly suffers, and every department of the mind becomes unhealthy. The man who, eager to become rich, takes time from his sleep for business purposes, draws from his brain capital.

278. *Regular and systematic mental exercise is essential to the health of nerve-tissue.* Exercise increases the flow of blood to the active part. We have seen this to be the case in the muscle, and that by use it is both enlarged and strengthened.

By what may the nervous system be enfeebled? Speak of the benefits of sleep and the amount needed. Observation. Why is mental exercise essential?

In like manner the nerve-tissue needs exercise, and as the gymnast becomes expert, not by spasmodic muscular efforts, but by accurate, persistent drill, so must the mental athlete gain his power by the regular performance of such exercise as he is able to bear.

279. *The amount of exercise should be adapted to the health and age of the individual.* If from any cause the nervous system be weakened, an amount of exercise which would be quite harmless to one in health may prove disastrous. The nerve-tissue of children and youth needs the same care as has been shown requisite for other tissues, and overwork that in the adult is followed by fatigue, easily removed by rest, in the child may result in irreparable injury.

280. *The required rest is often afforded by recreation and amusement.* Important as stated employment unquestionably is to the mental health, amusement or recreation is scarcely less so. Few persons, whatever their mental character or temperament, can safely dispense with these altogether. Even the most commanding intellects sometimes seek the recreation which their exhausting labors make necessary in forms of amusement which, to those who feel the necessity less, seem to be frivolous.

Observation 1.—To those whose life is one of severe toil and harassing care, amusements constitute almost the only practicable means for repairing the constant waste of the nervous energy. Especially is this want felt by women in the humbler walks of life, whose daily round of care and toil not only draws more largely than that of the stronger sex on the physical and mental energies, but is lightened by none of that relief which is afforded by a greater variety of duties and more frequent periods of rest.

2.—The brain, when severely taxed, is often rested by some kind of mental exercise, which, without being fatiguing, requires just enough effort to impart interest. Hence, a change from mathematics to the languages, or from these to music, poetry or painting, will give the needed relaxation.

281. *To maintain the highest mental vigor each faculty of the*

To what should the amount of exercise be adapted? Give the influence of recreation and amusement. Observations. What is essential to the highest mental vigor?

mind should receive its due share of cultivation. Our various faculties were not bestowed at random, to be used as inclination may prompt, but each has its appointed place in the mental economy. Each bears some relation to every other, making one harmonious whole. One must form habits of attention, accustom the mind to continuous thought, cultivate the reasoning powers and *beget a taste for exact knowledge,* if he would be in any measure equal to the intellectual effort essential to true success in every calling of life.

282. Man has also a *moral* faculty, the power of discriminating between right and wrong, which is quickly followed by the feeling of obligation to do the right and avoid the wrong. Upon the right use of these faculties depend the happiness and the destiny of man. The power of an approving conscience over the human mind, and consequently over the health of the Nervous System, cannot be over-estimated, while on the other hand the torments of an accusing conscience not only "cut the sinews of the soul's inherent strength," but snap one by one the gossamer filaments of the brittle thread of life.

§ 32. COMPARATIVE ANATOMY (Neurology).—*The Comparison of the Nervous System of other Mammals with that of Man—Of Birds—Of Reptiles—Of Amphibians—Of Fishes. Peculiar Arrangement of some Fishes. The Arrangement of the Nervous System of Mollusca—Of Radiata.*

283. Animals, whatever their structure may be, have certain relations with the external world; all nourish themselves; the lowest type, as the sponge, nourishes itself, as far as the result to itself is concerned, as does man. All Vertebrates do not possess a vertebral column, but all do possess something analogous to the spinal cord—a "*noto'-chord.*" The nervous system of Vertebrates is highly developed, and is made up of the brain and spinal cord. The latter are not represented in Invertebrates.

What is the moral faculty? Upon what depend the happiness and destiny of man? What is said of the Nervous System in Vertebrates and Invertebrates? Compare the Nervous System in other Mammals with that in man.

NERVOUS SYSTEM. 149

284. In other *Mammals*, the relative size of the cerebrum and cerebellum, except in the lowest order, as the Duck-mole, is about the same as in man; but the convolutions of the brain of other mammals are less developed than in man, and certain ganglions are comparatively larger. The brain of all mammals is formed on the same plan; in man alone the back lobe of the cerebrum overlaps the cerebellum. In the Horse and Ox the senses of smell, sight and hearing are acute, and their respective ganglions are large. In some animals, as the Mole, where vision is feeble, and in others where smell or hearing is obtuse, the ganglions are very small and the nerves very delicate.

285. In *Birds* the hemispheres are not united as in man; the cerebellum is proportionately larger than the medulla oblongata, and the comparative weight of the brain to the body is less than in mammals. The ganglions of sight in birds are large, which is particularly apparent in the Eagle, Vulture and Buzzard. In these, vision is not only far-reaching, but acute, and the same is true, to a certain extent, of smell and hearing. (Fig. 138.)

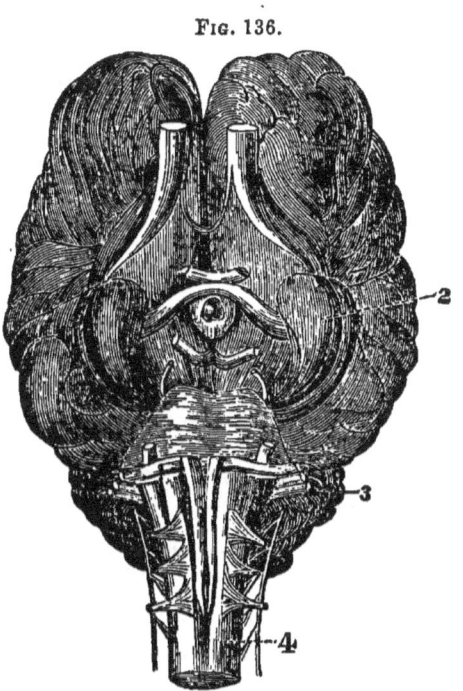

FIG. 136.

FIG. 136 (*Owen*). BASE OF BRAIN OF A HORSE.—1, Cerebrum. 2, Ganglion of sight. 3, Cerebellum. 4, Medulla Oblongata and Spinal Cord.

286. The brain of *Reptiles* is smooth and without convolutions. The hemispheres are hollow, and there is no striated

---

Describe the brain of Birds. Speak of the brain of Reptiles.

body. The cerebellum sends no prolongations across the medulla oblongata, as in mammals. The ganglions of sight and smell are, in general, large. Hearing is less complete than in mammals. (Fig. 137.)

287. In *Amphibians* the nervous system is but slightly developed. The cerebrum is small; the cerebellum is scarcely visible. (Fig. 137.)

288. The brain of the *Fish* is small; it does not fill the whole cranial cavity, there being found within it a spongy

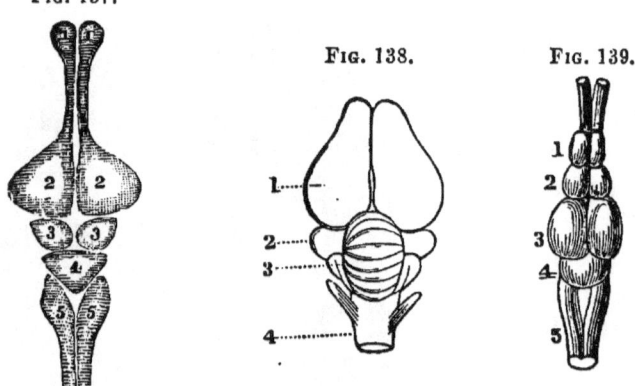

FIG. 137. BRAIN OF AN ALLIGATOR.—1, Olfactory ganglions. 2, Cerebrum. 3, Optic ganglions. 4, Cerebellum. 5, Medulla Oblongata and Spinal Cord.

FIG. 138. BRAIN OF A BIRD.—1, Cerebrum. 2, Optic ganglion. 3, Cerebellum. 4, Medulla Oblongata.

FIG. 139. BRAIN OF A FISH.—1, Olfactory ganglions. 2, Cerebrum. 3, Optic ganglions. 4, Cerebellum. 5, Medulla Oblongata and Spinal Cord.

fatty mass. The investment and protection of some of the organs of special sense are modified, as seen in the eye of the deep-sea shark, where one of the coats of the eye is bony, in order to protect this organ from the great pressure of the water. (Fig. 139.)

289. In the *Annulosa*, in general, each segment or ring has a pair of nervous ganglions. The ganglions of the nerves of special sensation are larger than those of general sensation.

290. The nervous system of insects is composed principally

Describe the Nervous System in Amphibians. Describe the brain of the Fish. Describe the Nervous System in the Annulosa.

of a double series of ganglions united by longitudinal cords. The brain ganglions are large, and give origin to the nerves of sight and the feelers (*antennæ*).

In the nervous system of the centipede, whose general structure is similar to that of other annulosa, the ganglions are arranged in pairs of nearly equal size, except the ganglion that answers to the brain, which is larger, along the under surface of the alimentary canal. Each pair is connected with the preceding, with the integument or skin, and with the muscles of its own ring, by sensitive and motor filaments of nerves.

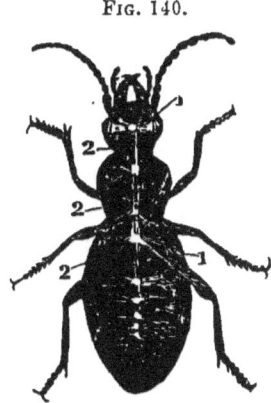

FIG. 140. DIAGRAM OF NERVOUS SYSTEM OF A BEETLE.—1, 1, Central ganglia. 2, 2, 2, Nerves that connect the ganglions.

291. In the *Mollusca* are found the ganglions and connective arrangement, with both sensitive and motor nerves, and on a plan corresponding to the body. (Fig. 142.)

292. In the RADIATA, the star-fish manifests one of the

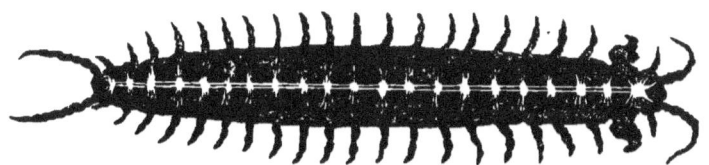

FIG. 141. DIAGRAM OF A CENTIPEDE.

simplest forms of the nervous system. It consists of a central mass, with five arms radiating from it. In the centre is the mouth, and beneath it the stomach or gastric cavity, which sends branches to each limb. The nervous system consists of five similar ganglions situated in the central portion at the base of the arms. These ganglions are connected, and

---

Describe the Nervous System in Insects. In the Centipede. Speak of the nervous system in Mollusca. Describe the nervous system in Radiata.

152 ANATOMY, PHYSIOLOGY AND HYGIENE.

each sends off nerve-filaments to the corresponding limbs. (Fig. 143.)

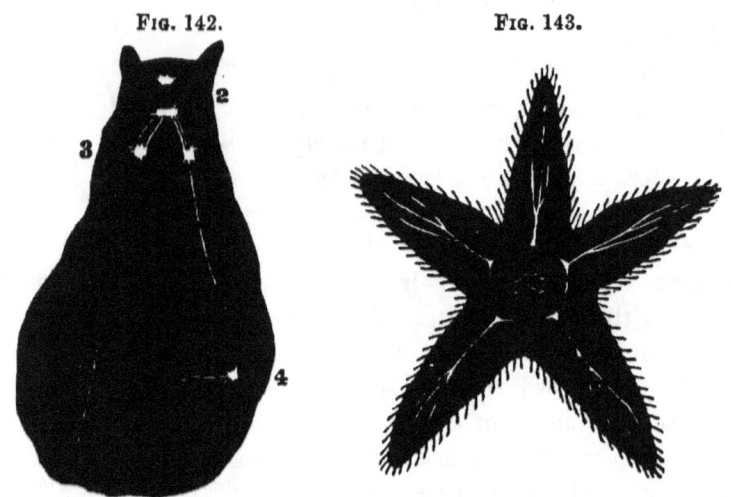

FIG. 142. DIAGRAM OF THE TYPE OF A MOLLUSCA.—1, Œsophagal ganglions. 2, Cerebral ganglions. 3, Pedal of locomotive ganglions. 4, Respiratory ganglions.
FIG. 143. DIAGRAM OF A RADIATA—THE STAR-FISH.

Observation.—The relations of the animal kingdom afford a striking evidence of divine unity, bound together in the closest harmony, and the work of Him who was the Beginning and will be the End.

Observation.

NERVOUS SYSTEM. 153

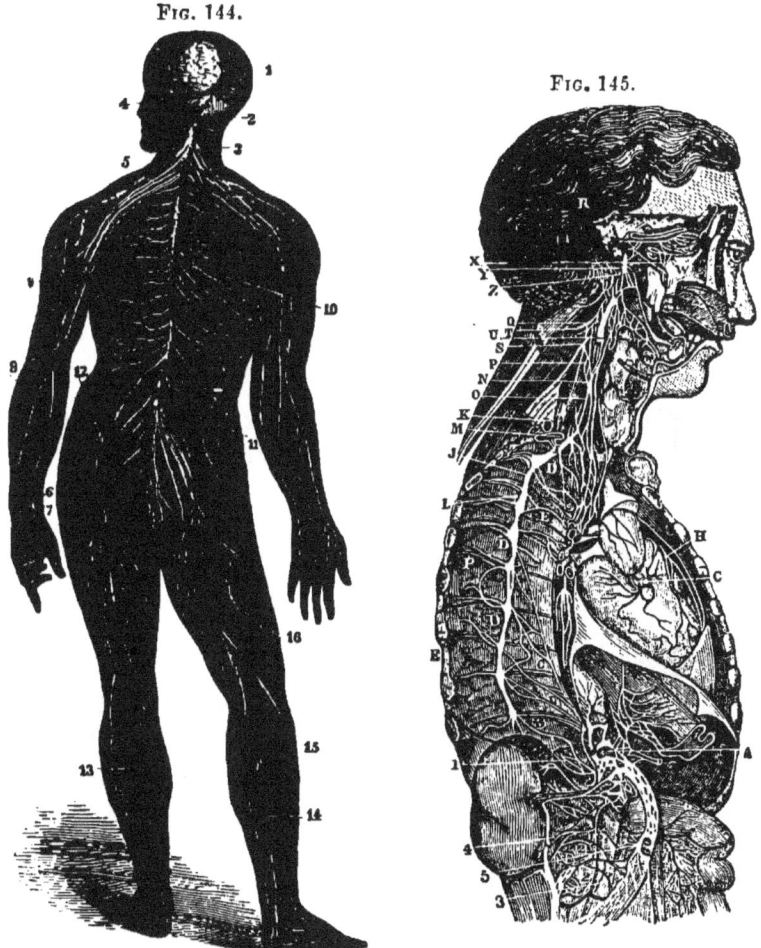

Fig. 144.

Fig. 145.

SYNTHETIC TOPICAL REVIEW
OF NERVOUS SYSTEM.
Sub-kingdoms, Divisions, Anatomy, Physiology, Hygiene.

State the Anatomy, the Physiology, the Hygiene, Human and Comparative

G*

CHAPTER XI.

SPECIAL SENSES.

293. As the Nervous System is the border land where the body touches the mind, we now shall speak of the bond of connection in the animal economy between the external and internal world—the *Special Senses*.

294. Our ideas of odor, of form, of sound and of taste are obtained from impressions made on the mind through the Senses. There are five senses—*Smell, Sight, Hearing, Taste* and *Touch*.

§ **33.** ANATOMY OF THE ORGANS OF SMELL.—*The Olfactory Nerves.*

295. The SENSE OF SMELL enables us to discern the odor or scent of substances. The special seat of the sense of smell is in the *delicate membrane* which lines the internal surface of the *Nose* and its passages. To this membrane the *Olfact'ory* or nerve of smell is distributed.

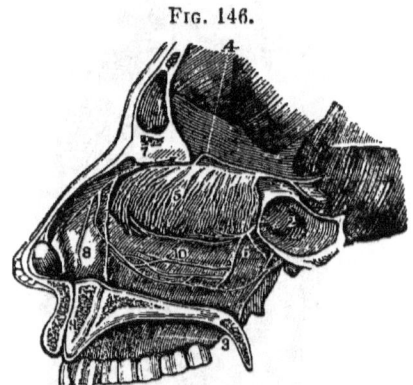

FIG. 146. A SIDE VIEW OF THE PASSAGE OF THE NOSTRILS.—4, The distribution of the first olfactory pair of nerves. 5, The fine divisions of this nerve on the membrane of the nose. 6, A branch of the fifth pair of nerves. 9, Upper jaw-bone and roof of the mouth.

296. To protect the delicate filaments of the nerve of smell, thus freely exposed to the air and to the painful stimulus of sharp, pungent odors, the membrane is kept constantly moist by a fluid secreted by the glands with which it is provided.

What is the bond of connection between the external and internal world in the animal economy? Name the Senses.

§ 34. Physiology of the Sense of Smell.—*Improved by Education.*

297. The SENSE OF SMELL aids man as well as the lower animals in selecting proper food. It also gives us pleasure by the inhalation of agreeable odors. When substances are presented to the nose the air that is passing through the nostrils brings the odoriferous particles of matter in contact with the filaments of the nerve of smell that are spread upon the membrane that lines the air passages, and the impression is then carried to the brain.

298. The sense of smell may be improved by cultivation. Thus the Indian can easily distinguish different tribes and different persons of the same tribe by the odor of their bodies. Next to touch, smell is of great importance to the blind.

299. This sense is seen to be remarkably acute in the Dog. He will trace his master's footsteps through thickly crowded streets and distinguish them from thousands of others; he will track the Hare over the ground for miles, guided only by the odor that it leaves in its flight. Some Fishes possess acuteness of smell in a remarkable degree, as shown in the nicety with which they select different baits.

§ 35. Hygiene of Smell.—*Perversion of the Sense of Smell.*

300. ACUTENESS OF SMELL requires that the brain and nerve of smell (olfactory) be healthy, and that the membrane that lines the nose be thin and moist. Any influence that diminishes the sensibility of the nervous filaments, thickens the membrane or renders it dry, impairs this sense.

Observation.—*Snuff,* when introduced into the nose, not only diminishes the sensibility of the nerve, but thickens the lining membrane. This thickening of the membrane obstructs the passage of air through the nostrils, and thus obliges "snuff-takers" to open their mouths when they breathe.

What is said of the Sense of Smell? What is the name of the nerve of Smell? How is this nerve protected? Give the function of Smell. Of what use is this sense? How susceptible of cultivation? Speak of the acuteness of this sense in the lower animals. On what does acuteness of smell depend? Give observation.

§ 36. ANATOMY OF THE ORGAN OF SIGHT.—*Different parts of the Eye.*

301. The SENSE OF SIGHT contributes more to the enjoyment and happiness of man than any of the other senses. By it we perceive the form, color, size and position of objects that surround us. The beautiful organ of sight (vision) is the *Eye*.

302. The EYE is shaped like a globe or ball, and is placed in a cavity in front of the skull. The sides of the globes are composed of three *Coats* (membranes). The interior of the globe is filled with certain substances called *Hu'mors*.

303. The COATS are three in number. 1st. The *Sclerot'ic* and *Corn'ea*. 2d. The *Cho'roid*, *I'ris* and *Cil'iary Processes*. 3d. The *Ret'ina*.

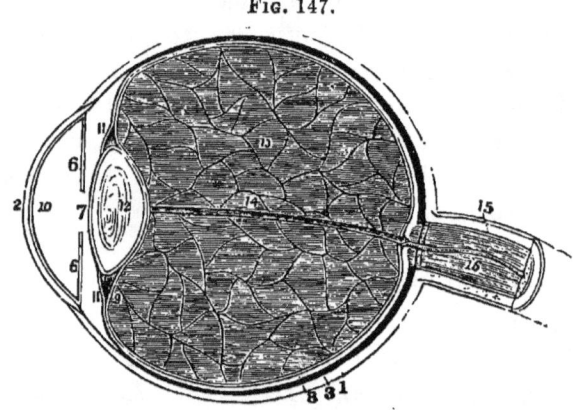

FIG. 147.

FIG. 147. A SECTION OF THE EYE.—1, The sclerotic coat. 2, The cornea. This connects with the sclerotic coat by a beveled edge. 3, The choroid coat. 6, 6, The iris. 7, The pupil. 8, The retina. 10, 11, 11, Chambers or cavities of the eye that contain the aqueous humor. 12, The crystalline lens. 13, The vitreous humor. 15, The optic nerve. 14, 16, An artery of the eye.

304. The HUMORS are also three in number. 1st. The *A'queous* (watery). 2d. The *Crys'talline* (lens). 3d. The *Vit'reous* (glassy). (Fig. 147.)

305. The SCLEROTIC coat is firm and its color white; hence it is frequently called the "white of the eye." From its

---

What sense contributes most to the enjoyment of man? What do we perceive by this sense? Give the structure of the Eye. Name the Coats. Name the Humors. Speak of the Sclerotic Coat.

SPECIAL SENSES. 157

toughness it forms the principal support to this organ. This membrane, with the cornea in front, encloses the eye.

306. The CORNEA is the transparent part of the eye in front, which projects more than the rest of the globe. It is shaped like the crystal of a watch, and in health gives the eye its sparkling brilliancy.

307. The CHOROID coat is of a dark color upon its inner surface. It contains a great number of blood-vessels, which give nourishment to different parts of the eye.

308. The IRIS is situated a short distance behind the cornea. It is the most delicate of all the muscles of the body. This part gives the blue, gray or black color to the eye. In the centre of the iris is an opening called the *pu'pil*, which enlarges or contracts according to the quantity of light that falls upon the eye.

309. On viewing the part of the eye near the pupil, small lines of a lighter color will be seen passing to the outer part of the iris; these are called CILIARY PROCESSES. They are about sixty in number. (Fig. 148.)

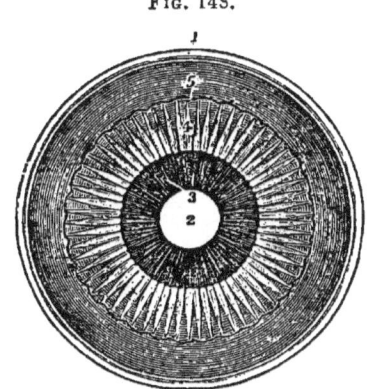

FIG. 148.

FIG. 148. A SECTION OF THE EYE SEEN FROM WITHIN.—1, The divided edge of the three coats. 2, The pupil. 3, The iris. 4, The ciliary processes. 5, The scolloped border of the retina.

310. The RETINA is the inner coat of the eye formed by the expansion of the optic nerve upon the inner side of the choroid coat, but not extending so far forward.

311. The AQUEOUS humor occupies the space between the cornea and crystalline humor both before and behind the iris.

312. The CRYSTALLINE humor (lens) lies behind the aqueous humor and pupil. Its form is different on the two

---

Speak of the Cornea. The Choroid Coat. Describe the Iris. The Retina. Speak of the Aqueous humor. Crystalline lens.

14

sides. When boiled it may be separated into layers like those of an onion.

Observation.—When the crystalline lens or the membrane which surrounds it is changed in structure so as to prevent the rays of light passing to the retina, the affection is called a *cataract*.

313. The VITREOUS humor is situated in the back part of the eye. It occupies more than two-thirds of the whole interior of the globe of the eye.

314. The OPTIC NERVE (or nerve of vision) extends from the brain to the back part of the eye, where it expands on a portion of the choroid coat. On this expansion the image of objects is first formed. (Fig. 149.)

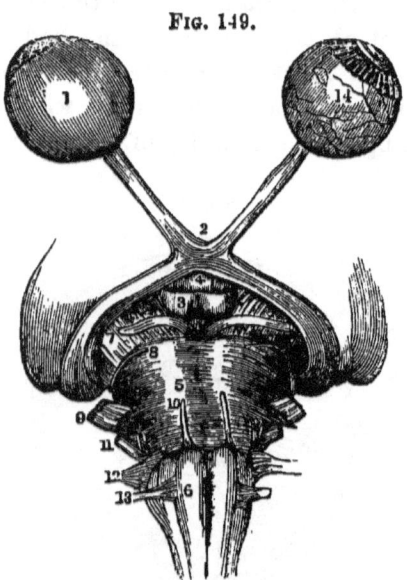

FIG. 149.

FIG. 149. THE OPTIC OR SECOND PAIR OF NERVES. -1, The eye-ball enclosed in its coats. 2, The crossing of the optic nerve. 3, 4, The brain. 5, 6, The commencement of the spinal cord. 7, 8, 9, 10, 11, 12, 13, The cranial nerves. 14, The globe with the sclerotic coat removed to show the retina.

315. The MUSCLES of the eye are six in number. They are attached at one extremity to the orbit behind the eye; at the other extremity they are inserted by broad, thin tendons to the sclerotic coat, near the junction of the cornea. The white, pearly appearance of the eye is caused by these tendons. (Fig. 150.)

Observation.—If the external muscle is too short, the eye is turned out, producing the "wall eye;" if the internal muscle is contracted, the eye is turned inward toward the nose, and is called a "cross eye."

316. The PROTECTING ORGANS of the eye are the *Orbits, Eyebrows, Eyelids* and *Lach'rymal* (tear) *Apparatus*. (Fig. 151.)

Give observation. Where is the Vitreous Humor situated? Describe the Optic Nerve. Speak of the Muscles of the Eye.

317. The ORBITS are deep, bony sockets in which the eyeballs are placed. The bottom of each orbit has a large opening giving passage to the optic nerve. These cavities are lined with a thick cushion of fat.

318. The EYEBROWS are the hairy arches forming the upper part of the boundary of the orbits.

319. The EYELIDS are two movable curtains having a delicate skin on the outside. Internally, they are lined by a

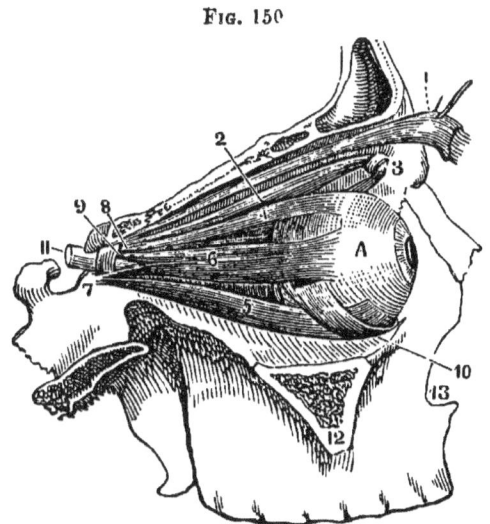

FIG. 150

FIG. 150. MUSCLES OF THE EYE.—A, The eye-ball. 1, 2, 4, 5, 6, 7, 8, 9 and 10, Muscles of the eye. 11, Optic nerve. 12, Cut surface of the cheek-bone. 3, The pulley arrangement through which the tendon of muscle 2 plays.

smooth membrane, called *conjunctiva*. The edges of the lids are furnished with hairs, called *eyelashes*.

320. The LACHRYMAL APPARATUS which secretes the tears consists of the *Lachrymal Gland* with its ducts, *Lachrymal Canals* and the *Nasal Duct*.

§ **37.** PHYSIOLOGY OF SIGHT.

321. The STRUCTURE OF THE EYE is beautifully adapted to the *laws of light*, a few of which it is necessary for us to

Name the protecting organs of the eye. Speak of the orbits of the eye. Describe the eyebrows. Give the structure of the eyelids. Describe the Lachrymal Apparatus State some of the laws of light.

notice. When light passes through a medium of unvarying density, the rays are in straight lines, but when it passes from a medium of one density into another of different density, they are *refracted* or bent from a straight course unless striking the medium perpendicularly, when they are unchanged. When light passes from one medium to another having a convex or concave surface instead of a flat surface, a great degree of refraction is produced, and the greater the curvature, the greater will be the amount of refraction.

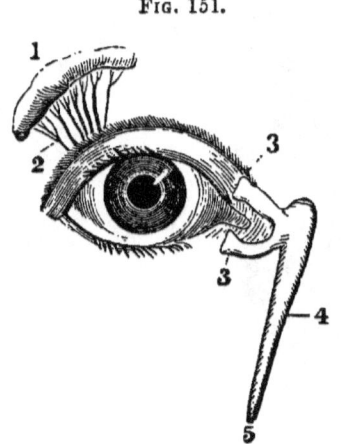

FIG. 151. VIEW OF LACHRYMAL GLAND AND NASAL DUCT.—1, The lachrymal gland. 2, Ducts leading from the lachrymal gland to the upper eyelid. 3, 3, The "tear points." 4, The nasal sac. 5, The termination of the nasal duct.

Illustration.—Fit a convex lens in an opening of the shutter of a darkened room; the rays of light will cross each other in the lens, and an inverted image of any object outside, as a tree or house, may be received upon a screen placed in the room at a certain distance behind the lens. The exact point where the image is most distinct is called the *focus of the lens*, and the distance from the lens to the image the *focus distance*. Now, in the eye the dark color of the choroid coat gives the darkened room, the retina the screen, the pupil is the opening in the shutter, and the three humors are the curved lenses. The rays of light from any object cross each other, and an inverted image is formed on the retina.

322. The shape of the CORNEA and the HUMORS of the eye seems admirably adapted to modify the rays of light in such a manner as to make the impression clear and distinct on the optic nerve.

Observation.—When the cornea or crystalline lens is too convex or the latter is too far from the retina, *short-sightedness* is produced, and the defect is measurably remedied by the use of concave glasses; when there is too little convexity, *long-sightedness* is the result, and convex glasses should be used. In old age, the humors being deficient in quantity, cause the flattening of the convex parts, hence the need of convex glasses. In the selection of glasses the lens for each eye should be chosen separately, as the foci of the two eyes do not usually exactly correspond; therefore a lens that will suit one eye may injure the other.

323. The SCLEROTIC COAT gives form to the eye, serves for the attachment of the muscles which move the eye in various directions.

324. The function of the dark color of the CHOROID COAT is to absorb all the luminous rays not necessary for sight.

325. The cushion of fat that lines the *orbits* enables the eye to move in all directions with freedom and without friction.

326. The EYEBROWS assist in shading the eyes when exposed to strong light, and they lend expression to some emotions of the mind.

327. The EYELIDS, by their movements, spread over the front of the eyeball a watery secretion, by which its surface is constantly bathed and its brilliancy and transparency kept unimpaired. Though the lids are closed by each *act of winking*, which is about six times a minute, the motion is so quick that it passes almost unobserved.

328. The EYELASHES so interlace that protection is given the eye from light substances floating in the air. They add very greatly to the expression of the eye.

329. The Horse is provided with a beautiful contrivance for protecting the eye. A cartilage is so arranged that it is made to sweep across the eye at the will of the animal, and removes dust or insects that may fall upon it.

In BIRDS and REPTILES a thin membrane is drawn across the eye, and so transparent as not to impair the sight. It gives protection to the eye from too strong light. Most insects are furnished with complex or many eyes.

Give the function of the Sclerotic coat. The Choroid coat. Speak of the Orbits of the Eyes. The Eyebrows. What is the use of the Eyelids? Of the Eyelashes? State the protection to the eye of the Horse. To other inferior animals.

§ 38. Hygiene of the Sense of Sight.

330. *The Eye is a delicate organ, requiring care to preserve it in health;* like other organs of the body, it should be exercised and then rested. The observance of this rule is particularly needful to those whose eyes are predisposed to inflammation. If the eye be used too long at one time, it becomes wearied and the power of sight diminished. On the contrary, if not called into exercise, its functions are enfeebled or permanently impaired.

331. *Using the eye in reading or writing,* also in looking at minute objects, is much more injurious when continued with an insufficient or a flickering light. It is poor economy to read habitually books too finely or badly printed. Sudden changes of light should be avoided. The want of sleep impairs this delicate and most valuable of all the senses.

§ 39. Anatomy of the Organ of Hearing.

332. Hearing, in utility, is scarcely inferior to that of sight. While we can see in only one direction at a time, we can hear from all directions. While the eye is useless in the dark, and veiled by its own curtains during sleep, the *Ear* is ever a faithful sentinel, warning us against danger.

333. The Ear is one of the most complicated organs in the human body. It is composed of three parts: 1st. The *External* ear. 2d. The *Middle* ear (Tym′panum). 3d. The *Internal* ear (Lab′yrinth).

334. The EXTERNAL ear presents many ridges and furrows, arising from the folds of the cartilage that forms it. A funnel-shaped tube extends from the external to the middle ear. The internal extremity of the tube is closed by a thin, semi-transparent membrane that separates the external from the middle ear; it is called the drum of the ear (Mem′brana Tym′pani). This and the bitter wax found around the hairs in the tube prevent insects from entering the head.

Give some rules for the preservation of the eye. Give the remarks relative to light and looking at minute objects. State the utility of hearing. What is said of the Ear? How divided? Give the structure of the External ear.

SPECIAL SENSES. 163

Observation.—Many animals have small muscles that move the external ear, in order to catch sounds from every direction. The hare, rabbit and horse afford good examples.

335. The MIDDLE ear is connected with the internal and most important cavity by four small bones, which are the most delicate and beautifully shaped bones in the body. These are so arranged as to form a chain from the drum of the ear to the labyrinth.

Observation.—From the middle ear a tube opens into the back part of the throat, called *Eustachian*, which admits air into this part of the ear. If this tube is closed by disease of the throat, hearing is impaired.

FIG. 152.

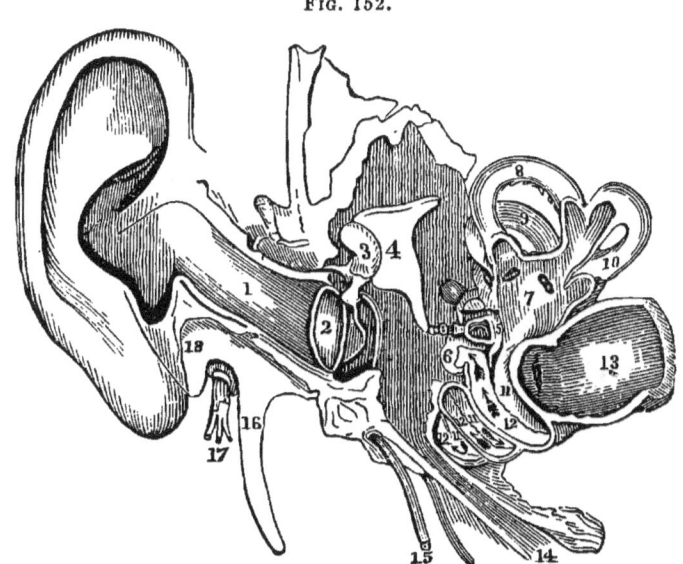

FIG. 152. A VIEW OF ALL THE PARTS OF THE EAR. 1, The tube that leads to the internal ear. 2, The membrana tympani. 3, 4, 5, The bones of the ear. 7, The central part of the labyrinth, named the vestibule. 8, 9, 10, The semicircular canals. 11, 12, The channels of the cochlea. 13, The auditory nerve. 14, The channel from the middle ear to the throat (Eustachian tube). 15, A nerve. 16, A process. 17, The seventh pair of nerves (facial). 18, A process of the temporal bone.

336. The INTERNAL ear is very intricate, and the uses of its various parts are not well known. It is called the labyrinth from its many windings. This part of the ear is com-

---

Describe the Middle ear. Observation. Speak of the Internal ear.

164 ANATOMY, PHYSIOLOGY AND HYGIENE.

posed of a three-cornered cavity called the *Ves'tibule*, the *Coch'lea* (from its resembling a snail shell), and the *semicir'cular* canals. The internal ear is the only part that is absolutely essential in hearing.

337. The VESTIBULE is never omitted. In the lower animals a simple sac which corresponds with this cavity is the whole organ of hearing. *Birds* have no external ear. In *Amphibians,* as the Frog, the tympanum can be seen back of

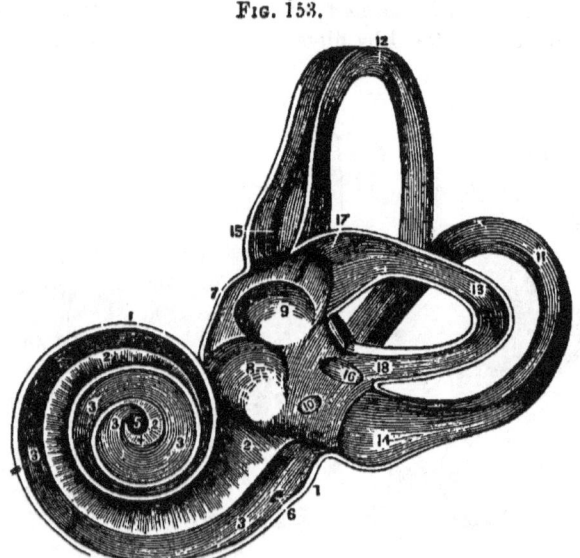

FIG. 153.

FIG. 153. A VIEW OF THE LABYRINTH LAID OPEN.—1, 1, The cochlea. 2, 2, 3, 3, Two channels that wind two and a half turns around a central point (5). 7, The central portion of the labyrinth, called the vestibule. 11, 12, 13, 14, 15, 16, 17, 18, The semicircular canals. Magnified.

the eye. In *Fishes* a semicircular canal is superadded to the tympanum.

§ 40. PHYSIOLOGY OF HEARING.

338. HEARING is that function by which we obtain a knowledge of the vibratory motion of bodies which constitute sound.

What part is absolutely essential in hearing? Speak of the organ of hearing in the lower animals. What is hearing?

SPECIAL SENSES.

339. The precise function of all the different parts of the ear are not known; that of the EXTERNAL EAR is to collect sounds and reflect them into the tube that connects the external with the middle ear. The drum (membrana tympani) receives all the impressions of the air which enter the tube and conveys them to the bones of the ear. It also serves to moderate the intensity of sound.

340. The supposed office of the MIDDLE EAR is to carry the vibrations made on the drum to the internal ear. This is effected by the air which it contains, and by the chain of small bones that are enclosed in this cavity.

341. But little is known of the functions of the INTERNAL EAR; its parts are filled with a watery fluid in which the filaments of the *auditory nerve* terminate.

342. The AUDITORY NERVE, like the optic, has but one function, that of special sensibility, or transmitting sound to the brain. The nerves which furnish the ear with ordinary sensibility proceed from the fifth pair.

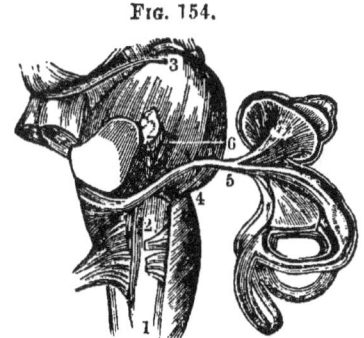

FIG. 154.

FIG. 154. A VIEW OF THE AUDITORY NERVE. —1, The spinal cord. 2, The medulla oblongata. 3, The lower part of the brain. 4, The auditory nerve. 5, A branch to the semicircular canals. 6, A branch to the cochlea.

343. The transmission of sound through the different parts of the ear will now be explained by the aid of Fig. 152. The vibrations of the air are collected by the External ear and conducted through the tube (1) to the drum (2). From the drum of the ear, the vibrations pass along the chain of bones (3, 4, 5). The bone (5) communicates with the Internal ear (7, 8, 9, 10, 11, 11, 11, 12, 12, 12). From the Internal ear the impression is carried to the brain by the Auditory nerve (13). (Figs. 153, 154.)

What is the function of the external ear? The middle ear? The internal ear? Speak of the nerve of hearing. Explain the transmission of sounds by Fig. 152.

§ 41. Hygiene of the Sense of Hearing.

344. *Acuteness in hearing* requires perfection in the structure and functions of the different parts of the ear and that portion of the brain from which the auditory nerve proceeds.

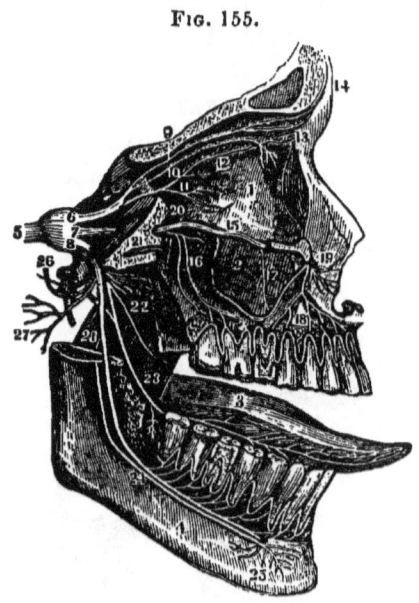

Fig. 155. The Distribution of the Fifth Pair of Nerves.—1, The orbit for the eye. 2, The upper jaw. 3, The tongue. 4, The lower jaw. 5, The fifth pair of nerves (*trigemini*). 6, The first branch of this nerve that passes to the eye. 9, 10, 11, 12, 13, 14, Divisions of this branch. 7, The second branch that passes to the teeth of the upper jaw. 15, 16, 17, 18, 19, 20, Divisions of this branch. 8, The third branch that is distributed to the tongue and teeth of the lower jaw. 23, The division of this branch (*gustatory*). 24, The division that is distributed to the teeth of the lower jaw.

Observation 1.—The common causes of impaired hearing are a thickening of the drum of the ear, an accumulation of wax upon its exterior surface, a closure of the Eustachian tube, disease of the brain, palsy of the auditory nerve and destruction of the middle and internal ear. It is injurious to put the heads of pins into the ear, as they frequently cause inflammation. The wax can be softened by dropping into the tube some oil, and in a few hours remove it by injecting warm soap-suds into the ear.

2.—When worms and insects find their way into the tube of the external ear, they can usually be driven out by dropping in warm olive oil.

345. *Hearing is improved by cultivation* from the habit of attention to the faintest impressions made upon the ear. Thus the skilled musician will detect the least discord in the blended effects of different instruments in a large orchestra.

By hearing we are enabled to appreciate the tone, the force and the direction of sounds, and also to gain informa-

---

Upon what does acuteness of hearing depend? Give some of the causes of impaired hearing. How can insects be removed from the ear? How is nicety of hearing acquired? How does hearing increase our happiness?

§ 42. ANATOMY OF THE ORGANS OF TASTE.

346. TASTE is the sense by which we perceive the flavor or relish of a thing. This sense differs from the other special senses, because it requires *actual contact* of the substances to be tasted, and these must be either naturally fluid or partially dissolved by the saliva.

347. The TONGUE is the principal organ of taste, though the sides of the cheeks and upper part of the throat share in this function. It is a muscular organ, and from the variety and arrangement of the muscles it is capable of various movements. If dry, hard substances are taken into the mouth, the tongue brings them in contact with the saliva to be dissolved. It not only aids in the mastication, but in the swallowing, of food.

The surface of the tongue is thickly studded with papillæ or points of various forms; these give this organ a velvety appearance. To these points the *Gust'atory*, or nerve of taste, is distributed. When fluids are taken into the mouth, the papillæ dilate and erect themselves, and the particular sensation excited is carried to the brain by this nerve. (Fig. 155.)

348. In all *Mammals* the tongue has nearly the same structure. In *Birds* it is generally cartilaginous and without nervous papillæ, so their taste cannot be acute. In some *Reptiles* the tongue is fleshy and large. In Serpents it is sharp and forked. In the Frog the tongue is darted out with great quickness of motion to catch the insects on which it feeds. The tongue of the Bee forms a little tube through which it sucks up the juices of flowers.

§ 43. PHYSIOLOGY OF THE SENSE OF TASTE.

349. The USE OF TASTE is to guide men and animals in

What is Taste? How does it differ from the other senses? Name the principal organ of Taste. Give its structure. Speak of the nerve of Taste. Speak of the tongue in Mammals. In Birds. In Reptiles. In the Bee. What is the use of Taste?

the selection of their food and to warn them against the introduction of injurious articles into the stomach.

This sense has been made to vary more than any other by the refinements of social life. It is modified by habit, and not unfrequently those articles which at first were disgusting become highly agreeable by persevering in the use of them, as in learning to chew tobacco and medicinal roots.

350. Substances which have an agreeable and healthy taste excite a flow of saliva, and digestion is better performed. The relish, however, diminishes as hunger is appeased. If indulgence of the appetite is continued beyond the necessities for food, nausea will compel the glutton to desist. The sympathy between the stomach and the sense of taste is a wise provision of the bountiful Giver.

Observation.—The *Sense of Taste becomes impaired* by the immoderate use of stimulants and condiments. These indulgences lessen the sensibility of the nerve. In children this sense is usually acute, and their preference is for food of the mildest character.

§ 44. Anatomy of the Sense of Touch.

351. Touch is the sense that enables us to tell whether a body is rough or smooth, cold or hot, sharp or blunt. This sense and feeling resides in the *nerves of the skin*. (P. 187.)

352. The nerves that contribute to the sense of touch proceed from the front half of the spinal cord. Where sensation is most acute we find the greatest number of nervous filaments.

Observation.—The sense of touch varies in different persons, and also in individuals of different ages. Thus the sensibilities of the child are more acute than those of the adult.

353. The sense of touch, though common to all parts of the *skin*, finds its highest development in the human hand. The delicacy of the structure of its covering, the different lengths of the fingers, with their pliancy, the opposing thumb,

---

How is Taste varied? How modified? How does Taste affect digestion? How is Taste impaired? What sense enables us to distinguish the qualities of bodies? Where does this sense reside? What connection between sensation and nervous filaments? What is said of the hand?

all contribute to make it an instrument of grace as well as an element of power.

354. In many *Mammals* the lips and tongue are the chief organs of touch. In the Elephant the finger-like projection at the end of the trunk possesses acute touch. In the Horse the lips are very sensitive. Fishes, on account of their scales, are nearly insensible.

355. This sense is modified by the condition of the brain and nerves, by the quantity and quality of the blood supplied to the skin, by the thickness of the cuticle, and by cultivation.

Observation.—Blind persons, by whom the beauties of the external world cannot be seen, cultivate this sense to such a degree that they can distinguish objects with great accuracy; and the rapidity with which they read books prepared for their use is a convincing proof of the niceness and extent to which the cultivation of this sense can be carried.

What is said of the sense of Touch in the lower animals? How is this sense modified? Give observation respecting the blind.

170 ANATOMY, PHYSIOLOGY AND HYGIENE.

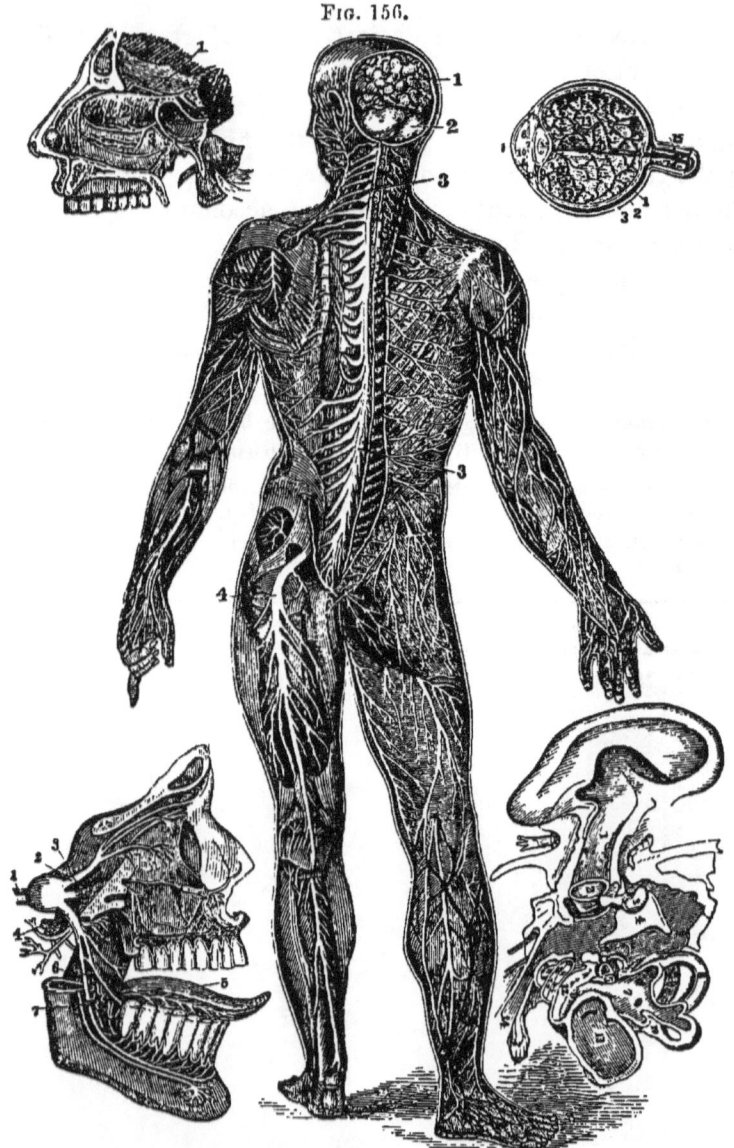

Fig. 156. A REPRESENTATION OF THE BRAIN, SPINAL CORD AND SPINAL NERVES.—1, The cerebrum. 2, The cerebellum. 3, 3, Spinal cord. 4, The sciatic nerve. A. DISTRIBUTION OF THE OLFACTORY NERVE.—1, 2, Nerve of smell. B. OPTIC NERVE.—15, The nerve of vision. C. THE GUSTATORY NERVE.—1, 2, 3, 4, Branches of the nerve of taste. D. AUDITORY NERVE.—13, Nerve of hearing.

SYNTHETIC TOPICAL REVIEW
OF THE SPECIAL SENSES.
Sense of Smell, Sight, Hearing, Taste and Touch.

State the Anatomy, Physiology and Hygiene of the Special Senses.

APPENDIX.

CHAPTER XII.

CARE OF THE SICK.*

§ **1.** In every home, however humble or dignified, woman is usually the Nurse. Nature seems to have endowed her in an especial manner to minister at the couch of disease and suffering. To be a good nurse requires a high type of womanhood; she should have both mental and physical power, blended with integrity and Christian trust.

If "good nursing is half the cure," how important that the daughter be early taught how to prepare drinks and nourishments, to administer medicine, and to perform the varied and important duties of the faithful nurse!

In the first stages of disease it is always proper treatment to *rest* both body and mind. It is wrong to tempt the appetite of a sick person; the disinclination for food is the warning of Nature that the system cannot well digest it.

The beneficial effects of *bathing* can hardly be over-estimated, but the mode of the bath should be directed by the medical adviser. The best time, however, for bathing, is when the patient feels most vigorous and freest from exhaustion. Care is necessary to wipe dry the skin, particularly between the fingers and toes, and also about the joints. Rubbing (friction) from a brush, moreen mitten or a dry flannel that has been saturated with salted water tends to relieve restlessness in patients. Chafing with the naked hand, making the movements *from* the brain and spine to the limbs, is peculiarly soothing, particularly if performed by a vig-

---

What is proper treatment in the first stages of disease? State some of the duties of the nurse in the sick-room.

* From the importance of the suggestions in this chapter, we would advise a thorough exercise weekly to the whole school, either orally or by recitation.

orous and healthy person. Air-baths have a tranquilizing influence.

The physician well knows that his attention to the sick is quite unavailing unless the nurse *obeys* his directions; for a nurse or immediate relatives or friends of the sick to put their judgment in opposition to that of the physician is not only arrogant, but endangers the patient. The *room* for the sick should be selected where sunlight may enter, and as far from external noise as possible. It is poor economy, not to say unkind, to keep a sick person in a small, ill-arranged bed-room, when a more spacious and airy room is kept for only occasional "callers." All superfluous furniture should be removed from the sick-room.

Quiet should reign in the sick-room. No more persons should enter or remain in it than the welfare of the patient demands. It is the duty of the physician to direct when visitors should be admitted or excluded, and the nurse should enforce the directions. The movements of the attendants should be gentle: no bustling to "clear up the room" at a fixed time; this should be done quietly, and when it will give the least annoyance to the sick. (It may be necessary to use a *damp* cloth in dusting the furniture, also the carpet, especially if the patient has disease of the lungs.) Creaking hinges should be oiled; shutting doors violently and heavy walking avoided. All unnecessary conversation should be deferred. If a colloquy must be carried on, let the tone be so high that the patient, if interested, can thoroughly comprehend it.

The *making of the bed* is often badly conducted. All bunches should be removed, the material of the bed laid even and a thin quilt spread smoothly over a mattress. When convenient, have the head of the bed northerly (62), and so situated, at least, that the sick man may look on something more pleasurable than a table of glasses and phials. A nurse should never manifest impatience in arranging the pillows, but try to adapt them to the comfort of the weary patient.

All utensils employed in the sick-room should be kept clean. Water designed for the patient to drink should not stand long in an open glass or pitcher, but be given fresh from a *spring* or well. A very sick person is fatigued by being raised to receive drinks; hence a bent tube or a cup with a spout should be used.

Both the *apparel and the bed-linen* should be changed more frequently in sickness than in health, and oftener in acute than in chronic diseases. All clothing, whether from the laundry or bureau, should be well dried and warmed by a fire previous to being put on the bed or the patient.

No agent is of more importance to the sick-room than *pure air;* hence the nurse with all convenient speed should remove everything that can emit an unpleasant odor. She should be chary of keeping ripe fruits or bouquets of flowers any length of time in the sick-chamber. When a disinfectant is needed, procure some at the druggist's. To change quickly and effectively the air of the sick-room, cover the patient's bed with an extra blanket and closely envelop his head and neck, except the mouth and nose; the door and windows can then be safely opened for a short time without detriment. After the windows are closed, retain the extra coverings on the patient until the room is of proper warmth. Unless duly protected, the patient should never feel *currents of air*, although *fresh air* should be constantly admitted into the sick room. (It is preferable to have pure air introduced from an adjoining apartment.) Few persons realize the *necessity* of fresh air being constantly admitted into an occupied room, whether by the healthy or the sick. The air exhaled from the lungs contains not only carbonic acid gas, but a *vapor* which gives the peculiar odor to the breath (159). All know its stifling character that have opened a close chamber that has been occupied during the night. Disease may be contracted by inhaling this vapor of respiration, as well as by actual contact with contagious matter.

A well-adjusted thermometer is indispensable, as the feelings of the patient or nurse are not to be relied on as a true index of the *temperature* of the room. Regulating the warmth of the patient is one of the many duties of the nurse. There is a "sweating temperature." The patient should no more be allowed to complain of *too much heat*, without an attempt at its reduction, than he should be permitted to remain chilly when the removal is possible.

The nurse should not confine herself to the sick-room longer than six hours at a time. She should exercise daily in the open air, also eat and sleep as regularly as possible. No doubts or fears of the patient's recovery, either by a look or a word, should be communicated by the nurse in the chamber of the sick; this duty devolves upon the physician.

Medicines assist the natural powers of the system to remove disease. They should be given regularly, judiciously and with a cheerful manner, and administered as directed by the physician. Life itself is often at the mercy of the nurse, and depends on the faithful discharge of her duty.

Drinks have a more decided influence upon the system than is generally admitted. They may be acid or alkaline, cold or hot, as the condition of the patient requires. The nurse should never depart from the quality of the drink, nor even exceed the due or prescribed quantity. Giving "herb-teas" without the sanction of the physician may cause serious evil.

In diseases of a typhoid character, and also in chronic ailments, where prostration from the waste of tissues and diminished generation of animal heat exists after the subsidence of active disease (when solid food cannot be taken), the gradual introduction into the system of the elements of food that is easily digested and assimilated becomes an important matter. These elements (the albuminous, saccharine and oleaginous substances of food, together with an increased amount of carbon) are found in the admixture of refined sugar with sweet pure milk and a small amount of pure alcoholic spirits in the form of "milk punch."

Solid food, as masticating beef steak or dry toasted crackers, is often preferable to gruels and other liquid food, especially when it is necessary to excite an action in the salivary and mucous glands. The food of the sick should be prepared in the neatest and most careful manner, and the nurse ought to obey implicitly the physician's directions about diet. When a patient is convalescent, the desire for food is generally strong; great care, firmness and patience is required that the food be prepared suitably and given at the proper time.

We append a few modes of preparing nourishment for the sick.

CRUST COFFEE.—Take light, sweet bread or crackers, and brown them *thoroughly* as you would coffee berry; when wanted for use, pour over boiling water (the crust will admit of several replenishings of boiling water); add sugar and cream to suit the condition of the patient.

GRUELS.—*Corn* meal requires to be boiled several hours to be suitable nourishment for the sick. The mode of preparing gruel should be suited to the case and directed by the physician. Wheat, or oat-meal, farina and sago, can be prepared in less time, though *they must be well cooked.* Add salt while cooking.

What is said of food and drinks? Name the means of nourishment, and tell how they may be prepared.

Egg Gruel.—Take the *yolks* of two eggs, boiled hard, and with a knife reduce them to a fine powder; beat this into a flour gruel made of new milk: salt and spices may be added if the condition of the patient admits.

BEEF TEA.—Meat contains principles that may be extracted, some by *cold*, others by *warm*, and others, again, by *boiling*, water; it should be cut very fine, and submitted for three hours each time, in succession, to half its weight of cold, of warm and of boiling water; the fluids strained from the first and second macerations are to be mixed with that strained from the boiling process, and the mixture should be brought to a boiling heat to cook it—the fat skimmed off; add a few drops of some acid, with salt, for a flavor.

A quicker, though less nutritious, mode of making beef tea, is to cut beef fine, put it in a glass bottle, cork it, place it in a kettle of cold water, then boil the beef from two to three hours; when cooled, strain the liquor and add salt.

§ **2.** The duty of the WATCHER is scarcely less responsible than that of the nurse, and, like the nurse, she should ever be cheerful, kind, firm and attentive in the presence of the patient.

The watcher should be prompt, and reach the house of the sick at an early hour; before entering the sick-room she should eat a simple, nutritious supper, and also during the night take some plain food. She should be furnished with an extra garment, as a heavy shawl, to wear toward morning, when the system becomes exhausted.

The directions about the sick, especially the administration of medicine, should be *written* for the temporary watcher. Whatever may be wanted during the night should be brought into the sick-chamber or the adjoining room before the family retires to sleep, that the slumbers of the patient be not disturbed by haste or searching for needed articles.

Sperm candles are preferable for the sick-room. Kerosene, in burning, emits a disagreeable odor, often annoying to the patient. All lights ought to be so arranged as not to shine or be reflected in the part of the room where the sick lie.

It is not necessary that watchers make themselves acceptable to the patient by exhausting conversation. If two watchers are needed, it is more imperative that they refrain from talking, and particularly *whispering*.

Give the duties of the Watcher. What directions about the administration of medicine? What lights are preferable for the sick-room?

Most sick persons have special need of nourishment about four or five o'clock in the morning.

The attendant upon the sick should not sit between the patient and the fire, and also should avoid sitting in the current of air that is flowing out of the room.

When taking care of the sick, light-colored clothing should be worn in preference to dark apparel, especially if the disease is of a contagious character. It is always safe for the watcher to change her apparel worn in the sick-chamber before entering upon her family duties. Disease is often communicated by the clothing.

It can hardly be expected that the farmer who has been laboring hard in the field, or the mechanic who has toiled during the day, is qualified to render all those little attentions that a sick person requires. Hence, would it not be more benevolent and economical to employ and *pay* watchers who are qualified by knowledge and *training* to perform this duty in a faithful manner, while the kindness and sympathy of friends may be *practically* manifested by assisting to defray the expenses of these qualified and useful assistants?

§ 3. THE TREATMENT OF WOUNDS OR INJURIES.—*Contusions* or *bruises* are generally treated by the injured person or some member of the family. The bruised limb should *rest*, be kept moderately warm, bathed frequently with tepid water and chafed moderately with the naked hand.

INCISED WOUNDS ("cuts").—At first there is free bleeding from the many divided capillaries. If no large vein or artery is severed, the flow of blood will soon cease; press the gaping wound together, and trickle on cold water until the blood and all foreign matter is removed; then apply narrow strips of adhesive plaster.

The union of the divided parts is effected by the action of the blood-vessels, and not by "healing salves" or "ointments." The only object of the dressing is to keep the parts together and protect the wound from air and impurities. Nature performs her own cure. Such wounds seldom need a second dressing, and should not be opened till the incisions are healed. To lessen the liability of a reopening, a proper position for the union should be regarded. If the wound be between the knee and ankle, and on the front part, extend the knee and bend up the ankle; if on the back part,

---

What caution is necessary in sitting by the sick? Give the manner of dressing wounds—Bruises—Cuts.

APPENDIX. 177

reverse the movement, and, in general, suit the position to the case.

LACERATED WOUNDS.—In these injuries, the jagged, torn parts do not heal by the "first intention," but "matter" is formed before healing. Cleanse the parts with cold water and apply a soft poultice. All wounds made by blunted or pointed instruments, as nails, should be examined by a surgeon.

Wounds from *Poisonous Serpents* or *Rabid Animals* should have cupping-glasses immediately applied, or sucked by the mouth. Give freely alcoholic stimulants until a physician arrives.

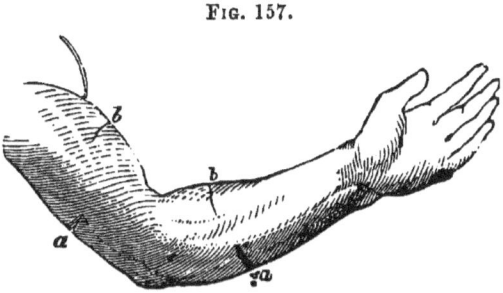

FIG. 157. *a, a,* REPRESENTATION OF WOUNDS on the back part of the fore-arm. *b, b,* Wounds of the anterior part of the arm and fore-arm. By bending the elbow and wrist, the incisions at *a, a,* are opened, while those at *b, b,* are closed. Were the arm extended at the elbow and wrist, the wounds at *a, a,* would be closed, and those at *b, b,* would be opened.

Observation.—Although animal poisons, when introduced into the circulating fluid through the broken surface of the skin, frequently cause death, yet they can be taken into the mouth and stomach with comparative impunity; as when the mucous membrane which lines these parts is not broken, poisons are rarely absorbed.

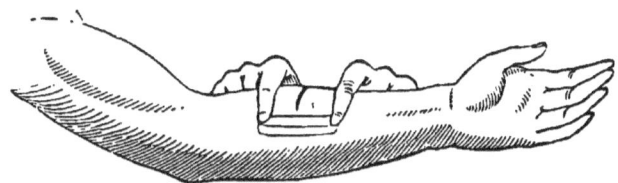

FIG. 158 REPRESENTS THE MANNER of applying adhesive strips to wounds.

HEMORRHAGE FROM DIVIDED ARTERIES SHOULD BE ARRESTED, otherwise the heart soon ceases its action, and the person faints. If a large artery is wounded, every beat of the pulse throws out the blood in jerks. Until surgical help can be sum-

---

Give the treatment of lacerated wounds. Wounds from rabid animals.

moned, the flow of the blood may be stopped either by compressing the vessel between the wound and the heart, or by compressing the end of the artery next the heart in the wound.

After compression as described and illustrated, take a square piece of cloth, or handkerchief, twist it cornerwise, and tie a hard knot in the middle. Place the knot over the artery between the wound and the heart, carry the ends around the limb and tie loosely. Place a stick under the handkerchief near the last tie, and twist till the fingers can be removed from the compression without a return of the bleeding. When an artery in a limb be cut, elevate the limb as far as possible, till the bleeding ceases.

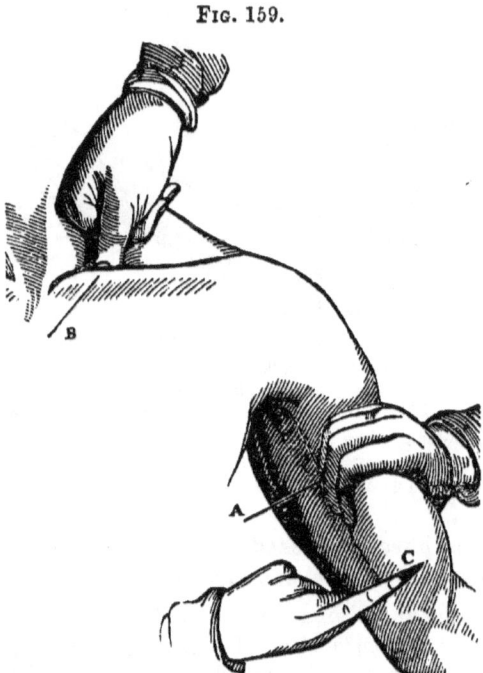

FIG. 159.

FIG. 159. THE MANNER OF COMPRESSING DIVIDED ARTERIES.—A, Compressing the large artery of the arm with the thumb. B, The subclavian artery. C, Compressing the divided extremity of an artery *in the wound* with a finger.

BURNS AND SCALDS.—When blisters are formed, the outer skin is separated from the other layer or cuticle by the effusion of serum, or water; this fluid should be let free by puncturing the cuticle, care being taken not to remove the thin raised skin, as it makes the best possible protection to the sensitive, inflamed tissues beneath. When this thin outside layer of skin is removed, immediately cover the denuded parts with wheat flour, or a plaster made of lard and bees'-wax or the white of an egg; in a word, substitute a cuticle to protect the exposed nerves from the air. When dressings are applied, they should not be removed until they become dry and irritating.

How may hemorrhage be arrested? Speak of Burns and Scalds, and their treatment.

To prevent the formation of blisters when only a small patch of the skin is scalded or burned, apply *steadily* cold water until the smarting pain ceases; then put on a simple dressing, "not to take out the fire or to heal it," but to protect the injured membrane.

When the skin in particular spots is exposed to excessive pressure or friction, it becomes too much thickened, producing "*Corns*." These are not necessarily confined to the feet, but are produced in front of the clavicle of the soldier from the pressure of his musket, or on the knee of the cobbler. The pain of the callosity is due to its exciting inflammation in the sensitive dermis upon which it

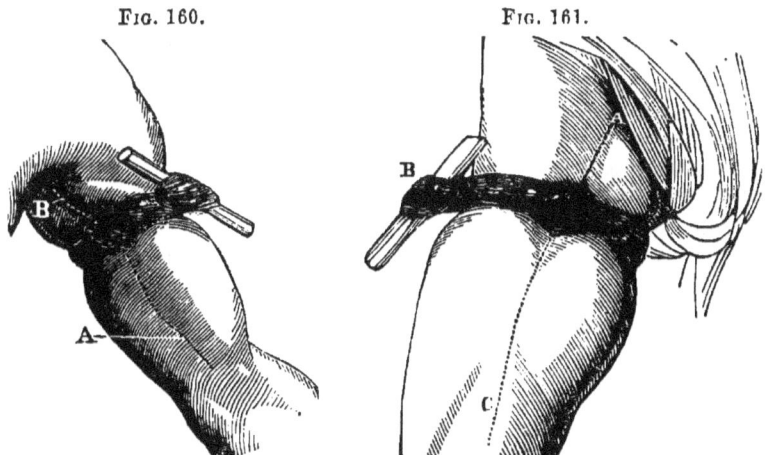

Fig. 160. THE METHOD OF APPLYING THE KNOTTED HANDKERCHIEF, to compress a divided artery. A, B, Track of the brachial artery.

Fig. 161. A, C, The track of the femoral artery; the compress applied near the groin

presses. Remove the pressure, and the affected part is restored to its healthy state.

FROST-BITE is usually manifested first upon parts unprotected by covering, as the face or ears, and especially the nose. In such case the skin first becomes red, from congestion of the dilated capillary vessels; next it becomes bluish, from arrest of the circulation; and afterward of a dead white hue. To restore circulation and sensibility rub the frozen part with snow or apply iced water. Keep the sufferer at first in a cold room, and let the return to a higher temperature be *gradual* and cautious, or *gangrene* may supervene.

How may the formation of blisters be avoided in slight burns? How are "Corns" produced? Speak of Frost-bite.

The CHILBLAIN is not produced by the action of cold, but by the effect of heat on the chilled extremity. Bathe the inflamed parts with a mixture of sweet oil, glycerine and camphorated spirits.

§ 4. *Asphyxia—Treatment of Apparently Drowned Persons.*

"DROWNING.—1st. Treat the patient *instantly on the spot, in the open air,* freely exposing the face, neck and chest to the breeze, except in severe weather. 2d. In order to *clear the throat,* place the patient gently on the face, with one wrist under the forehead, that all fluid, and the tongue itself, may fall forward and leave the entrance into the trachea or windpipe free. 3d. To excite respiration or *breathing,* turn the patient slightly on his side and apply some irritating or stimulating agent to the nostrils, as hartshorn or dilute ammonia, cologne, etc. 4th. Make the face warm by brisk friction, then dash cold water upon it. If not successful, lose no time to *imitate respiration.*

"ARTIFICIAL RESPIRATION.—First, place the patient on the face and turn the body gently but *completely on the side and slightly beyond;* then again on the face, repeating alternately these movements deliberately and perseveringly *fifteen times* only in a minute.

"*Observation* 1.—When the face position is resumed, make a uniform and efficient pressure along the *spinal column* (back-bone), removing the pressure immediately before turning on the side; continue these measures. (The pressure increases the expiration, and rotation commences inspiration.)

"2.—When the patient lies on the chest, this cavity is compressed by the weight of the body, and *expiration* takes place; when turned on the side, this pressure is removed, and *inspiration* occurs.

"3.—Rub the limbs *upward* with *firm pressure* and with energy, to aid the return of venous blood to the heart.

"4.—Rub the body briskly till it is dry and warm, then dash *cold* water upon it and repeat the rubbing.

"Avoid the immediate removal of the patient, as it involves a *dangerous loss of time.* Avoid the warm bath. Substitute for the patient's wet clothing, if possible, such other covering as can be *instantly* procured, each bystander supplying a coat until flannel blankets are obtained. To excite inspiration. let the surface of the body be *slapped briskly* from time to time with the hand."

(*From Marshall Hall's Treatment of Asphyxia from Drowning, Chloroform or Gas.*)

How are Chilblains produced? How may asphyxiated persons be recovered? State the method of Artificial Respiration.

APPENDIX.

POISONS AND THEIR ANTIDOTES.

§ 5. Every mother or housekeeper should know and have at hand some available remedies for the common poisons which are frequently taken either by accident or design. Nearly every poison has its antidote, which, *if used at once*, may prevent much suffering and even death.

When known that poison has been taken into the stomach, the first thing is to evacuate it by the use of the stomach-pump or an emetic, unless vomiting takes place spontaneously.

As an emetic, ground *Mustard* mixed in warm water is always safe. Take one tablespoonful to one pint of warm water. Give the patient one-half in the first instance and the remainder in fifteen minutes, if vomiting has not commenced. In the interval drink copious draughts of warm water. Irritate the throat with a feather or the finger, to induce vomiting. After vomiting has begun, give mucilaginous drinks, such as flaxseed tea, gum-arabic water, or slippery elm.

If the patient is drowsy, give a strong infusion of cold coffee, keep him walking, slap smartly on the back, use electricity; it may be well to dash cold water on the head to keep the patient awake. After the poison is evacuated from the stomach, to sustain vital action give warm water and wine or brandy. If the limbs are cold, apply warmth and friction.

In ALL cases of poisoning call immediately a physician, as the after-treatment is of great importance.

POISONS.	ANTIDOTES OR REMEDIES FOR POISONS.
Aconite (Monkshood). Belladonna (Deadly Nightshade). Bryony. Camphor. Conium } (Water Hemlock). Cicuta } Croton Oil. Digitalis (Foxglove). Dulcamara (Bitter-Sweet). Gamboge. Hyoscyamus (Henbane). Laudanum.	For *Vegetable* poisons give an emetic of *Mustard;* drink freely of warm water; irritate the throat with a feather to induce vomiting. Keep the patient awake until a physician arrives.

When poisons have been taken, what is to be done? Name the most common poisons and their antidotes.

APPENDIX.

POISONS.	ANTIDOTES OR REMEDIES FOR POISONS.
Lobelia. Morphine. Opium. Paregoric. Sanguinaria (Blood-Root). Savin Oil. Spigelia (Carolina Pink). Stramonium (Thorn Apple). Strychnine (Nux Vomica). Tobacco.	For *Vegetable* poisons give an emetic of *Mustard;* drink freely of warm water; irritate the throat with a feather to induce vomiting. Keep the patient awake until a physician arrives.
Arnica.	*Vinegar* and water.
Prussic Acid. Bitter Almonds (Oil of). Laurel Water.	Drink, at once, one teaspoonful of *Water of Hartshorn* (ammonia) in one pint of water.
Ammonia (Hartshorn). Potash. Soda.	Antidote is *Vinegar* or *Lemon Juice*, followed with sweet, castor or linseed oil. Thick cream is a substitute for oil.
Iodine.	Starch or wheat flour beat in water. Take a Mustard emetic.
Saltpetre (Nitrate of Potassa). Chili Saltpetre (Nitrate of Soda).	Take, at once, a *Mustard* emetic; drink copious draughts of warm water, followed with oil or cream.
Lunar Caustic (Nitrate of Silver).	Two teaspoonfuls of table salt (chloride of sodium) mixed in one pint of water.
Corrosive Sublimate (bug poison). White Precipitate. Red Precipitate. Vermilion.	Beat the *Whites of six Eggs* in one quart of cold water; give a cupful every two minutes, to induce vomiting. A substitute for white of eggs is *soap-suds* slightly thickened with wheat flour. Emetics should only be given by a physician.

APPENDIX.

POISONS.	ANTIDOTES OR REMEDIES FOR POISONS.
Arsenic. Cobalt (fly powder). King's Yellow. Ratsbane. Scheele's Green.	Use a *stomach-pump* as quickly as possible, or give a Mustard emetic until one is obtained. After free vomiting, give large quantities of *Calcined Magnesia*. The antidote for Arsenic is *Hydrated Peroxide of Iron*.
Acetate of Lead (Sugar of Lead). White Lead. Litharge.	Use a *Mustard* emetic, followed by Epsom or Glauber Salts. The antidote is diluted *Sulphuric Acid*.
Antimony (Wine of). Tartar Emetic.	The antidote is ground *Nutgall*. A substitute, oak or Peruvian bark, followed by a teaspoonful of paregoric.
Pearl-ash. Ley (from wood-ashes). Salts of Tartar.	Drink freely of *Vinegar* and water, followed with a mucilage, as flaxseed tea.
Sulphuric Acid (Oil of Vitriol). Nitric " (Aquafortis). Muriatic " (Marine). Oxalic "	Drink largely of water or a mucilage. It is important that something *be given quickly* to neutralize the acid. The antidote is *Calcined Magnesia*. Chalk, lime, strong soap-suds, are substitutes for magnesia.
Matches (Phosphorus). Rat Exterminator.	Give two tablespoonfuls of *Calcined Magnesia*, followed by mucilaginous drinks.
Verdigris. Blue Vitriol.	The antidote is *Cooking Soda* or *White of Eggs*. Drink milk freely.
Sting of Insects.	*Ammonia*, or cooking soda moistened with water, applied in the form of a paste. The wound may be sucked, followed by applications of water.
Tainted Crabs, Oysters or Fish.	Use a Mustard emetic, and drink freely of vinegar and water.
Charcoal Fumes. Gas or Burning Fluid.	*Fresh* air and Artificial Respiration.

SUMMARY—SYNTHETIC TOPICAL REVIEW.

Sect. 1. Definitions.	} Chap. I. *General Remarks.* }		
" 2. Anatomy of.			
" 3. Physiology of.	} Chap. II. *The Bones.*	} Division I. *Motory Apparatus.*	
" 4. Hygiene of.			
" 5. Comparative Osteology.			
" 6. Anatomy of.	} Chap. III. *The Muscles.*		
" 7. Physiology of.			
" 8. Hygiene of.			
" 9. Comparative Myology.			
" 10. Anatomy of.	} Chap. IV. *The Digestive Organs.*		
" 11. Physiology of.			
" 12. Hygiene of.			
" 13. Comparative Splanchnology			
" 14. Anatomy of.	} Chap. V. *The Absorbents.*		
" 15. Physiology of.			
" 16. Hygiene of.			
" 17. Anatomy of.	} Chap. VI. *The Organs of Respiration.*	} Division II. *Nutritive Apparatus.*	} Mammals.
" 18. Physiology of.			
" 19. Hygiene of.			
" 20. Comparative Pneumonology.			
" 21. Anatomy of.	} Chap. VII. *The Skin.*		
" 22. Physiology of.			
" 23. Hygiene of.			
" 24. The Blood, anatomy of.	} Chap. VIII. *The Circulation.*		
" 25. Physiology of.			
" 26. Hygiene of.			
" 27. Comparative Angiology.			
" 28. Assimilation, Secondary.	} Chap. IX. *Assimilation.*		
" 29. Anatomy of.	} Chap. X. *The Nervous System.*	} Division III. *Nervous Apparatus.*	
" 30. Physiology of.			
" 31. Hygiene of.			
" 32. Comparative Neurology.			
" 33. } " 34. } Smell. " 35. }	} Chap. XI. *The Organs of Special Sense.*		
" 36. } " 37. } Sight. " 38. }			
" 39. } " 40. } Hearing. " 41. }			
" 42. } " 43. } Taste. " 44. }			
" 1. Care of the Sick.	} Chap. XII. *Appendix.*		
" 2. Watchers, duty of.			
" 3. Treatment of Wounds, Hemorrhage and Burns.			
" 4. Asphyxia.			
" 5. Poisons and Antidotes.			

State the Anatomy, the Physiology and the Hygiene of Mammals

GLOSSARY.

Ab-do'men. [L. *abdo*, to hide.] That part of the body which lies between the thorax and the bottom of the pelvis.

Ab-o-ma'sum. [L.] The fourth stomach of a ruminant animal.

Ab-sorp'tion. [L. *ab*, and *sorbeo*, to suck up.] The imbibition of a fluid by an animal membrane or tissue.

A-chil'lis. A term applied to the tendon of the two large muscles of the leg.

Al-bu'men. [L. *albus*, white.] An animal substance of the same nature as the white of an egg.

Al've-o-lar. [L. *alveolus*, a socket.] Pertaining to the sockets of the teeth.

Am-phib'i-ans. [Gr. αμφι, *amphi*, both, and βιος, *bios*, life.] A class of animals so formed as to live on land and in water. At one period of their existence they breathe by gills, at another by lungs.

A-nas'to-mose. [Gr.] The communication of arteries and veins with each other.

An-a-tom'i-cal. Relating to the parts of the body when dissected or separated.

A-nat'o-my. [Gr. ανα, *ana*, through, and τομη, *tomē*, a cutting.] The description of the structure of animals. The word *anatomy* properly signifies dissection.

An-gi-ol'o-gy. [Gr. αγγειον, *angeion*, a vessel, and λογος, *logos*, discourse.] A description of the vessels of the body, as the veins and arteries.

An-i-mal'cu-le. [L. *animalcula*, a little animal.] Animals that are only perceptible by means of a microscope.

An-nu-lo'sa. [L. *annulus*, a ring.] Furnished with rings. The same as the articulate animals.

An-ten'næ. [L. *antennæ*, a yard-arm.] The jointed horns or feelers possessed by the articulata or annulosa.

An-te'ri-or. [L.] Before or in front in place; opposed to posterior.

A-ort'a. [Gr. αορτη, *aortē*; from αηρ, *aēr*, air, and τηρεω, *tēreō*, to keep.] The great artery that arises from the left ventricle of the heart.

Ap-pa-ra'tus. [L. *apparo*, to prepare.] An assemblage of organs designed to produce certain results.

Ap-pend'ix. [L. *ad*, and *pendeo*, to hang from.] Something appended or added.

A'que-ous. [L. *aqua*, water.] Partaking of the nature of water.

A-rach'noid. [Gr. αραχνη, *arachnē*, a spider, and ειδος, *eidos*, form.] Resembling a spider's web; a thin membrane that covers the brain.

Ar'bor. [L.] *Arbor vitæ*. The tree of life. A term applied to a part of the brain.

Ar'te-ry. [Gr. αηρ, *aēr*, air, and τηρεω, *tēreō*, to keep; because the ancients thought that the arteries contained only air.] A tube through which blood flows *from* the heart.

A-ryt-e'noid. [Gr. αρυταινα, *arutaina*, a ewer, and ειδος, *eidos*, form.] The name of a cartilage of the larynx.

As-phyx'i-a. [Gr. α, *a*, not, and σφυξις, *sphyxis*, pulse.] Originally, want of pulse; now used for suspended respiration or apparent death.

As-sim-i-la'tion. [L. *ad* and *similis*.] The conversion of nutriment into the fluid or solid substance of the body.

As-trag'a-lus. [Gr.] The name of a bone of the foot; one of the tarsal bones.

Aud-it-o'ri-us. [L. *audio*, to hear.] Pertaining to the organ of hearing.

Au'ri-cle. [L. *auricula*, the external ear from *auris*, the ear.] A cavity of the heart; the "deaf ear."

A-ZOTE'. [Gr. α, a, not, and ζωη zōē, life.] Nitrogen. One of the constituent elements of the atmosphere; so named because it will not sustain life.

BI'CEPS. [L. bis, twice, and caput, a head.] A name applied to muscles with two heads at one extremity.

BI-CUS'PIDS. [L. bis, two, and cuspis, a point.] Teeth that have two points upon their crown.

BILE. [L. bilis.] A viscid, bitter fluid secreted by the liver.

BI-PEN'NI-FORM. [L. bis, two, and penna, a feather.] Having fibres on each side of a common tendon.

BRACH'I-AL. [L. brachium.] Belonging to the arm.

BRAIN. [Sax. bragen.] The soft mass or viscus enclosed within the skull-bones.

BRANCH'I-Æ. [L.] Gills; a term applied to the organs of respiration in certain animals which live in water.

BRONCH'I-A, -Æ. [L.] A division of the trachea that passes to the lungs.

BRONCH-I'TIS. [L.] An inflammation of the bronchia.

BUR'SÆ MU-CO'SÆ. [L. bursa, a purse, and mucosa, viscous.] Small sacs containing a viscid fluid, situated about the joints, under tendons.

CÆ'CUM, -A. [L.] Blind; the name given to the commencement of the colon.

CAL'CIS. [L.] The heel-bone.

CAP'IL-LA-RY. [L. capillus, a hair.] Resembling a hair; a small tube.

CAP'SULE. [L. capsula, a little chest.] A membranous bag enclosing a part.

CAR'BON. [L. carbo, a coal.] Pure charcoal. An elementary combustible substance.

CAR-BON'IC. Pertaining to carbon.

CAR'DI-AC. [Gr. καρδια, kardia, heart.] Relating to the heart, or upper orifice of the stomach.

CAR'NE-A, -Æ. [L. caro, carnis, flesh.] Fleshy.

CAR-NIV'O-ROUS. [L. caro, flesh, and voro, to eat.] Eating or feeding on flesh.

CA-ROT'ID. [Gr. καρος, karos, lethargy.] The great arteries of the neck that convey blood to the head. The ancients supposed drowsiness to be seated in these arteries.

CAR'PUS, -I. [L.] The wrist.

CAR'TI-LAGE. [L. cartilago.] Gristle. A smooth, elastic substance, softer than bone.

CAU'DAL. [L.] Pertaining to a tail.

CA'VA. [L.] Hollow. Vena Cava, a name given to the two great veins of the body.

CELL. [L.] A little bag or bladder containing fluid or other matter.

CEL'LU-LAR. [L. cellula, a little cell.] Composed of cells.

CER-E-BEL'LUM. [L.] The hinder and lower part of the brain, or the little brain.

CER'E-BRO-SPI'NAL. Relating to the brain and spine.

CER'E-BRUM. [L.] The front and large part of the brain. The term is sometimes applied to the whole contents of the cranium.

CER'VI-CAL. [L. cervix, the neck.] Relating to the neck.

CHEM'IS-TRY. [Gr. kimia, hidden art.] It relates to those operations by which the intimate nature of bodies is changed, or by which they acquire new properties.

CHEST. [Sax.] The thorax; the trunk of the body from the neck to the abdomen.

CHOR'DA, -Æ. [L.] A cord; an assemblage of fibres.

CHO'ROID. [Gr. χοριον, chorion.] A term applied to several parts of the body that resemble the skin.

CHYLE. [Gr. χυλος, chulos, juice.] A nutritive fluid, of a whitish appearance, which is extracted from food by the action of the digestive organs.

CHYME. [Gr. χυμος, chumos, juice.] A kind of grayish pulp formed from the food in the stomach.

CIL'IA-RY. [L. cilia, eyelashes.] Belonging to the eyelids.

CLAV'I-CLE. [L. clavis, a key.] The collar-bone; so called from its resemblance in shape to an ancient key.

CLO-A'CA. [L. a sink.] The common cavity into which open the intestinal canal and the ducts of other organs in some of the inferior animals.

COC'CYX. [Gr.] An assemblage of bones joined to the sacrum.

COCH'LE-A. [Gr. κοχλω, kochlō, to twist;

or L. *cochlea*, a screw.] A cavity of the ear resembling in form a snail-shell.

Co'LON. [Gr. κωλον, *kōlon*, I arrest.] A portion of the large intestine.

CO-LUM'NA, -Æ. [L.] A column or pillar.

COM'MIS-SURE. [L. *committo*, I join together.] A point of union between two parts.

CON-JUNC-TI'VA. [L. *con*, together, and *jungo*, to join.] The membrane that covers the anterior part of the globe of the eye.

COR-A'COID. [Gr. κοραξ, *korax*, a crow, and ειδος, *eidos*, form.] A process of the scapula shaped like the beak of a crow.

CO'RI-UM. [Gr. χοριον, *chorion*, skin.] The true skin.

CORN'E-A. [L. *cornu*, a horn.] The transparent membrane in the fore part of the eye.

COR'PUS CAL'LO-SUM. [L. *corpus*, a body, and *callus*, hard.] The great band of nervous matter which unites the two hemispheres of the brain.

COS'TA. [L. *costa*, a coast or side.] A rib.

CRI'COID. [Gr. κρικος, *krikos*, a ring, and ειδος, *eidos*, form.] A name given to a cartilage of the larynx from its form.

CRUS-TA'CE-A. [L.] One of the classes of the annulosa, including lobsters, crabs, etc.

CRYS'TAL-LINE. [L. *crystallinus*, consisting of crystal.] *Crystalline lens*, one of the humors of the eye.

CU'BOID. [Gr. κυβος, *kubos*, a cube, and ειδος, *eidos*, form.] Having nearly the form of a cube.

CU-NE'I-FORM. [L. *cuneus*, a wedge.] The name of bones in the wrist and foot.

CUS'PID. [L. *cuspis*, a point.] Having one point.

CU-TA'NE-OUS. [L. *cutis*, skin.] Belonging to the skin.

CU'TI-CLE. [L. *cutis*, skin.] The external layer of the skin.

CU'TIS VE'RA. [L. *cutis*, skin, and *vera*, true.] The internal layer of the skin; the true skin.

DE'CUS-SA'TION. [L. *decutio*, I divide.] A union in the shape of an X or cross.

DEL'TOID. [Gr. δελτα, *delta*, the Greek letter Δ, and ειδος, *eidos*, form.] The name of a muscle that resembles in form the Greek letter Δ.

DEN'TAL. [L. *dens*, tooth.] Pertaining to the tooth.

DE-PRESS'OR. [L.] The name of a muscle that draws down the part to which it is attached.

DERM'IS. [Gr. δερμα, skin.] The natural covering of animal bodies.

DE-SCEND'ENS. [L. *de* and *scando*, to climb.] Descending, falling.

DI'A-PHRAGM. [Gr. διαφραγμα, *diaphragma*, a partition.] A muscle separating the chest from the abdomen; the midriff.

DI-GES'TION. [L. *digestio*.] The process of dissolving food in the stomach and preparing it for circulation and nourishment.

DOR'SAL. [L. *dorsum*, the back.] Pertaining to the back.

DU-O-DE'NUM. [L. *duodenus*, of twelve fingers' breadth.] The first portion of the small intestine.

DU'RA MA'TER. [L. *durus*, hard, and *mater*, mother.] The outermost membrane of the brain.

DYS-PEP'SI-A. [Gr. δυς, *dūs*, bad, and πεπτω, *pepto*, to digest.] Indigestion or difficulty of digestion.

EN-AM'EL. [Fr.] The smooth, hard substance which covers the crown or visible part of a tooth.

EP-I-DERM'IS. [Gr. επι, *epi*, upon, and δερμα, *derma*, the skin.] The superficial layer of the skin.

EP-I-GLOT'TIS. [Gr. επι, *epi*, upon, and γλωττα, *glōtta*, the tongue.] One of the cartilages of the glottis; during the act of swallowing it prevents the food entering the larynx.

EP-I-THE'LI-UM. [Gr. επι, *epi*, upon, and θηλη, *thělě*, a nipple.] A layer of soft cells covering the surface of the lining membranes and part of the skin.

ETH'MOID. [Gr. ηθμος, *ěthmos*, a sieve, and ειδος, *eidos*, a form.] A bone of the skull.

EU-STA'CHI-AN TUBE. A channel from the fauces to the middle ear; named from Eustachi, who first described it.

EX'CRE-MENT. [L. *excerno*, to separate.] Matter excreted and ejected; alvine discharges.

EX'CRE-TO-RY. A little duct or vessel, destined to receive secreted fluids and to excrete or discharge them; also a secretory vessel.

EX-HA'LANT. [L. *exhalo*, to send forth vapor.] Having the quality of exhaling or evaporating.

EX-TEN'SOR. [L.] A name applied to a muscle that serves to extend any part of the body; opposed to *Flexor*.

FA'CIAL. [L. *facies*, face.] Pertaining to the face.

FALX. [L. *falx*, a scythe.] A process of the dura mater shaped like a scythe.

FAS'CI-A. [L. *facia*, a band.] A tendinous expansion or aponeurosis.

FAS-CIC'U-LUS, -LI. [L. *fascis*, a bundle.] A little bundle.

FAUX, -CES. [L.] The top of the throat.

FEM'O-RAL. Pertaining to the femur.

FE'MUR. [L.] The thigh-bone.

FE-NES'TRA, -UM. [L. *fenestra*, a window.] A term applied to some openings into the internal ear.

FIBRE, FIB-RIL'LÆ. } [L. *fibra*.] An organic filament or thread which enters into the composition of every animal and vegetable texture.

FI'BRIN. A peculiar organic substance found in animals and vegetables; it is a solid substance, tough, elastic, and composed of thready fibres.

FI'BRO-CAR'TI-LAGE. An organic tissue, partaking of the nature of fibrous tissue and that of cartilage.

FIB'U-LA. [L. a clasp.] The outer and lesser bone of the leg.

FIL'A-MENT. [L. *filamenta*, threads.] A fine thread, of which flesh, nerves, skin, etc. are composed.

FLEX'ION. [L. *flectio*.] The act of bending.

FOL'LI-CLE. [L. *folliculus*, a small bag.] A little bag or sac formed of an animal membrane; the orifice is generally minute.

FORE'ARM. The part of the upper extremity between the elbow and hand.

FUNC'TION. [L. *fungor*, to perform.] The action of an organ or system of organs.

FUR'CU-LA. [L. *furca*, a fork.] The V-shaped bone, or wish-bone of birds, formed by united clavicles.

GAN'GLI-ON, -A. [Gr. γαγγλιον, *ganglion*, a knot.] An enlargement in the course of a nerve.

GAS'TRIC. [Gr. γαστηρ, *gastēr*, the stomach.] Belonging to the stomach.

GAS-TROC-NE'MI-US. [Gr. γαστηρ, *gaster*, the stomach, and κνημη, *knēmē*, the leg.] The name of large muscles of the leg which serve to draw the heel upward.

GEL'A-TIN. [F. *gelo*, to conceal.] A concrete animal substance, transparent and soluble in water.

GLAND. An organ consisting of tubes and follicles, with blood-vessels interwoven from which the gland elaborates its secretion.

GLO'BULE. [L.] A small particle of matter of a spheroidal or round form. The red particles which swim in the watery part of the blood.

GLOS'SA. [Gr.] The tongue. Names compounded with this word are applied to muscles of the tongue.

GLOT'TIS. [Gr.] The narrow opening at the upper part of the larynx.

GLU'TE-US, -I. [Gr.] A name given to muscles of the hip.

GUS-TA'TO-RY. [L. *gusto*, to taste.] A name given to the nerve of taste.

HEART. [Sax.] The muscular organ which is the primary organ of the blood's motion in an animal body.

HEM'OR-RHAGE. [Gr. αἱμα, *haima*, blood, and ῥηγνυω, *rēgnuō*, to burst.] A discharge of blood from an artery or vein.

HERB-IV'O-ROUS. [L. *herba* and *voro*.] Feeding on herbs or vegetables.

HE-PAT'IC. [Gr. ἡπαρ, *hēpar*.] The liver.

HU'MER-US. [L.] The bone of the arm.

HY'DRA. [Gr. ὑδρα, *hudra*, a water-serpent.] The fresh-water polypes.

HY'DRO-GEN. [Gr. ὑδωρ, *hydōr*, water, and γενναω, *gennaō*, to generate.] A gas which constitutes one of the elements of water.

HY'GI-ENE. [Gr. ὑγιεινον, *hugieinon*, health.] The part of medicine which treats of the preservation of health.

HY'OID. [Gr. υ and ειδος, *eidos*, shape.] A bone of the tongue resembling the Greek letter Upsilon in shape.

HY'PO-GLOS'SAL. Under the tongue. The name of a nerve of the tongue.

IL'E-UM. [Gr. ειλω, eilō, to wind.] A portion of the small intestines.

IN-CI'SOR. [L. incido, to cut.] A front tooth that cuts or divides.

IN'DEX. [L. indico, to show.] The forefinger; the pointing finger.

IN-FU-SO'RI-A. [L.] A class of Protozoa; so called from their frequent occurrence in organic infusions.

IN-GLU'VIES. [L.] The crop. The first stomach of birds.

IN-NOM-I-NA'TA, -TUM. [L. in, not, and nomen, name.] Parts which have no proper name.

IN-SEC-TIV'O-RA. [L. insectum, an insect, voro, I devour.] Living upon insects.

IN-TER-COST'AL. [L. inter, between, and costa, a rib.] Between the ribs.

IN-TER-MAX'IL-LA-RY. [L. inter, between, maxilla, the jaw-bone.] Being between the cheek-bones.

IN-TER-VERT'E-BRAL. [L.] Between the vertebra.

IN-TES'TINES. [L. intus, within.] The canal that extends from the right orifice of the stomach to the anus; about thirty feet long.

I'RIS. [L., the rainbow.] The colored circle that surrounds the pupil of the eye.

I'VO-RY. A hard, solid, fine-grained substance of a fine white color; the tusk of an elephant.

JE-JU'NUM. [L., empty.] A portion of the small intestine.

JU'GU-LAR. [L. jugulum, the neck.] Relating to the throat; the great veins of the neck.

LA'BI-I. [L.] The lips.

LAB'Y-RINTH. [Gr.] The internal ear; so named from its many windings.

LACH'RY-MAL. [L. lachryma, a tear.] Pertaining to tears.

LAC'TE-AL. [L. lac, milk.] A small tube of animal bodies for conveying chyle from the intestine to the thoracic duct.

LAR-YNX. [Gr. λαρυγξ. larunx.] The upper part of the windpipe.

LA-TIS'SI-MUS, -MI. [L., superlative of latus, broad.] A term applied to some muscles.

LE-VA'TOR. [L. levo, to raise.] A name applied to a muscle that raises some part.

LIG'A-MENT. [L. ligo, to bind.] A strong, compact substance serving to bind one bone to another.

LI'VER. The largest gland in the system. It is situated below the diaphragm, and secretes the bile.

LOBE. A round projecting part of an organ.

LOB'ULE. A division of a glandular organ communicating with a single duct.

LUM'BAR. [L. lumbus, the loins.] Pertaining to the loins.

LYMPH. [L. lympha, water.] A colorless fluid in animal bodies, and contained in vessels called lymphatics.

LYM-PHAT'IC. A vessel of animal bodies that contains or conveys lymph.

MA'JOR. [L., greater.] Greater in extent or quantity.

MAM'MALS, -IA. [L.] Include man and all the ordinary quadrupeds.

MAR'ROW. [Sax.] A soft, oleaginous substance contained in the cavities of bones.

MAS-SE'TER. [Gr. μασσαομαι, massaomai, to chew.] The name of a muscle of the face.

MAS'TI-CATE, MAS-TI-CA'TION. [L. mastico.] To chew; the act of chewing.

MAS'TOID. [Gr. μαστος, mastos, breast, and ειδος, eidos, form.] The name of a process of the temporal bone behind the ear.

MAX-IL'LA. [L.] The jaw-bone.

ME-A'TUS. [L. meo, to go.] A passage or channel.

ME-DI-AS-TI'NUM. A membrane that separates the chest into two parts.

ME'DI-UM, -A. [L.] The space or substance through which a body passes to any point.

MED'UL-LA-RY. [L. medulla, marrow.] Pertaining to marrow.

ME-DUL'LA OB-LON-GA'TA. Commencement of the spinal cord.

ME-DUL'LA SPI-NA'LIS. The spinal cord.

MEM'BRA-NA. A membrane; a thin, white flexible skin formed by fibres interwoven like network.

MES'EN-TER-Y. [Gr. μεσος, mesos, in the midst, and εντερον, enteron, the intestine.] The membrane by which the intestines are attached to the spinal column.

ME-TA-CAR'PUS. [Gr. μετα, meta, after or beyond, and καρπος, karpos, wrist.] The part of the hand between the wrist and fingers.

ME-TA-TAR'SUS. [Gr. μετα, *meta*, after or beyond, and ταρσος, *tarsos*, the tarsus.] The instep.

MID'RIFF. [Sax. *mid* and *hrife*, the belly.] See DIAPHRAGM.

MI'NOR. [L.] Less, smaller. A term applied to several muscles.

MI'TRAL. [L. *mitra*, a mitre.] The name of the valves on the left side of the heart.

MO'LAR. [L. *mola*, a mill.] The name of some of the large teeth.

MOL LUS'CA. [L. *mollis*, soft.] One of the sub-kingdoms, comprising species whose bodies are soft and inarticulate. The digestive and other organs are enclosed by a fleshy sac.

MO'TOR, -ESS. [L. *moveo*, to move.] Exciting motion. A term applied to certain nerves.

MO-TO'RY. [L.] Giving motion.

MU'COUS. Pertaining to mucus.

MU'CUS. A viscid fluid secreted by the mucous membrane.

MUS'CLE. A bundle of fibres enclosed in a sheath. The lean meat of animals.

MUS'CU-LAR. Pertaining to a muscle.

MY-OL'O-GY. [Gr. μυς, *mus*, a muscle, and λογος, *logos*, a discourse.] A description of the muscles.

NA'SAL. Relating to the nose.

NERVE. [L. *neures*, a string.] White cords arising from the brain or spinal cord. An organ of sensation and motion in animals.

NERV'OUS CEN'TRE. A collection of gray nervous matter which receives impressions and originates the nervous impulses.

NEU-RI-LEM'MA. [Gr. νευρον, *neuron*, a nerve, and λεμμα, *lemma*, a sheath.] The sheath or covering of a nerve.

NEU-ROL'O-GY. [Gr. νευρον, *neuron*, a nerve, and λογος, *logos*, a discourse.] A description of the nerves of the body.

NI'TRO-GEN. That element of the air which is called azote.

NOR'MAL. [L. *norma*, a rule.] Conforming to the ordinary standard.

NO'TO-CHORD. [Gr. νοτος, *notos*, the back, χορδη, *chordē*, a string.] A cellular rod which is developed in the embryo vertebrates immediately beneath the spinal cord, and which is usually replaced in the adult by the spinal column.

NU-CLE-O'LUS. [L.] The minute solid particle found in the interior of the nucleus of some cells.

NU'CLE-US. [L. *nux*, a nut.] The central part of any body, or that about which matter is collected.

NU-TRI'TION. The art or process of promoting the growth or repairing the waste of the system.

Œ-SOPH'A-GUS. [Gr. οιω, *oiō*, to carry, and φαγω, *phagō*, to eat.] The name of the passage through which the food passes from the mouth to the stomach.

OL-FACT'O-RY. [L. *oleo*, to smell, and *facio*, to make.] Pertaining to smelling.

O-MA'SUM. [L.] The third stomach of Ruminants.

O-MEN'TUM. [L.] The caul. A fold of the peritoneum.

OM-NIV'O-ROUS. [L. *omnia*, everything, *voro*, I devour.] Feeding indiscriminately upon all kinds of food.

OP'TIC. [Gr. οπτομαι, *optomai*, to see.] Relating to the eye.

OR-BIC'U-LAR. [L. *orbis*, a circle.] Circular.

OR'GAN. A part of the system destined to exercise some particular function.

OS. [L.] A bone; the mouth of anything.

OS'SE-OUS. Pertaining to bones.

OS'SI-FY. [L. *ossa*, bones, and *facio*, to make.] To convert into bone.

OS-TE-OL'O-GY. [Gr. οστεον, *osteon*, a bone, and λογος, *logos*, a discourse.] The part of anatomy which treats of bones.

O-VIP'A-ROUS. [L. *ovum*, an egg, *pario*, I bring forth.] Producing young from eggs.

OX'Y-GEN. A permanently elastic fluid, invisible and inodorous. One of the components of atmospheric air.

PAL'ATE. [L.] The roof of the mouth.

PAN'CRE-AS. [Gr. παν, *pan*, all, and κρεας, *kreas*, flesh.] A long gland situated near the stomach (in domestic animals called the "sweet-bread").

PA-PIL'LA, -Æ. [L.] Small conical prominences. They constitute the roughness of the upper surface of the tongue.

PA-RI'E-TAL. [L. *paries*, a wall.] A bone of the skull.

PA-ROT'ID. [Gr. παρα, *para*, near, and ωτος, *ōtos*, the gen. of ους, *ous*, ear, the ear.] The name of the largest salivary gland.

PA-TELL'A, -Æ. [L.] The knee-pan.

PA-THET'I-CUS, -CI. [Gr. παθος, *pathos*, passion.] The name of the fourth pair of nerves.

PEC'TO-RAL. [L.] Pertaining to the chest.

PE-DUN'CLE. [L. *pes*, the foot.] A name applied to parts of the brain from the resemblance to a flower-stalk.

PEL'VIS. [L.] The basin formed by the large bones at the lower part of the abdomen.

PEN'NI-FORM. [L. *penna*, a feather.] Having the form of a feather or quill.

PER-I-CAR'DI-UM. [Gr. περι, *peri*, around, and καρδια, *kardia*, the heart.] A membrane that encloses the heart.

PER-I-CRA'NI-UM. [Gr. περι, *peri*, around, and κρανιον, *kranion*, the cranium.] A membrane that invests the skull.

PER-I-OS'TE-UM. [Gr. περι, *peri*, around, and οστεον, *osteom*, a bone.] A membrane that immediately invests the bones of animals.

PER-I-STAL'TIC. [Gr. περιστελλω, *peristellō*, to involve.] A movement like the crawling of a worm.

PER'MA-NENT. Durable; lasting.

PER-SPI-RA'TION. [L. *per*, through, and *spiro*, to breathe.] The excretion from the skin.

PHAL'ANX, GES. [Gr. φαλαγξ, *phalanx*, an army.] Three rows of small bones forming the fingers or toes.

PHAR'YNX. [Gr. φαρυγξ, *pharunx*.] The upper part of the œsophagus.

PHYS-I-OL'O-GY. [Gr. φυσις, *phusis*, nature, and λογος, *logos*, a discourse. The science of the functions of the organs of animals and plants.

PI'A MA'TER. [L. good mother.] The name of one of the membranes of the brain.

PLA-TYS'MA. [Gr. πλατυς, *platūs*, broad.] A muscle of the neck.

PLEU'RA, -Æ. [Gr. πλευρα, *pleura*, the side.] A thin membrane that covers the inside of the thorax and also forms the exterior coat of the lungs.

PLEX'US. [L. *plecto*, to weave together.] Any union of nerves, vessels or fibres in the form of network.

PNEU-MO-GAS'TRIC. [Gr. πνευμων, *pneumōn*, the lungs, and γαςτηρ, *gastēr*, the stomach.] Belonging to both the stomach and lungs.

PNEU-MO-NOL'O-GY. [Gr. πνευμων, *pneumōn*, the lungs, and λογος, *logos*, a discourse.] A description of the lungs.

POL'YPE. [Gr. πολυς, *polus*, many, πους, *pous*, foot.] An aquatic animal of the radiate type.

PONS. [L.] A bridge. *Pons varolii*, a transverse band of nervous fibres passing in a curved form from one side of the cerebellum to the other, spanning the medulla oblongata.

PON'TI-O DU'RA. [L., hard portion.] The facial nerve; seventh pair.

POR'TI-O MOL'LIS. [L., soft portion.] The auditory nerve; seventh pair.

POS-TE'RI-OR. [L. *post*, after.] Opposed to anterior.

PRE-HEN'SION. [L.] Grasping; seizing.

PRO-NA'TOR. [L. *pronus*, turned downward.] The muscle of the forearm that moves the palm of the hand downward.

PRO-TO-ZO'A. [Gr. πρωτος, *prōtos*, first, and ζωη, *zoōn*, animal.] The Infusoria or lowest class of animals.

PRO-VEN-TRIC'U-LUS. The second or true stomach in birds.

PUL-MON'IC, } [L. *pulmo*, the lungs.]
PUL'MO-NA-RY. } Belonging or relating to the lungs.

PU'PIL. A little aperture in the centre of the iris through which the rays of light pass to the retina.

PY-LO'RUS. [Gr. πυλωρος, *pulōros*, a gatekeeper.] The lower orifice of the stomach, with which the duodenum connects.

RA'DI-US. The name of one of the bones of the forearm.

RA-DI-A'TA. [L. *radio*, to shoot rays.] Includes those animals whose parts are arranged round an axis and on one or several radii.

RA'DI-ATE. Having lines or fibres that diverge from a point.

REC'TUM. [L.] Straight. The third and last portion of the intestines.

RE'FLEX AC'TION. An involuntary action of the nervous system by which an external impression, conducted by a sensory nerve, is reflected or converted into a motor impulse.

REG'I-MEN. [L. *rego*, to govern.] The systematic regulation of the food and drink.

REP'TILES, -IA. [L. *repo*, to creep.] A class of animals that breathe air from birth and are generally covered with scales.

RE-SID'U-UM. [L.] Waste matter. The fæces.

RES-PI-RA'TION. [L. *re*, again, and *spiro*, to breathe.] The act of breathing; inspiring air into the lungs and expelling it again.

RE-SPI'RA-TO-RY. Pertaining to respiration; serving for respiration.

RE-TIC'U-LUM. [L. *rete*, a net.] The second stomach of ruminants.

RET'I-NA. [L. *rete*, a net.] The essential organ of sight. One of the coats of the eye, formed by the expansion of the optic nerve.

RO-DEN'TI-A. [L. *rodo*, to gnaw.] A class of mammals having two large cutting teeth in each jaw, separated from the molar teeth by an empty space.

RO-TUN'DUM, -A. [L.] Round; circular.

RU'GA, -Æ. [L.] A wrinkle; a fold.

RU'MI-NANT. [L.] An animal that chews the cud.

SA'CRUM. [L., sacred.] The bone which forms the posterior part of the pelvis and is a continuation of the spinal column.

SA-LI'VA. [L.] The fluid which is secreted by glands and poured into the mouth. It is a solvent of the food.

SAL'IV-A-RY. Pertaining to saliva.

SAN-GUIN'E-OUS. [L. *sanguis*, the blood.] Abounding with blood; plethoric.

SAR-TO'RI-US. [L. *sartor*, a tailor.] A term applied to a muscle of the thigh.

SCA-LE'NUS. [Gr. σκαληνος, *skalēnos*, unequal.] A term applied to some muscles of the neck.

SCAPH'OID. [Gr. σκαφη, *skaphē*, a little boat.] The name applied to one of the wrist-bones.

SCAP'U-LA. [L.] The shoulder-blade.

SCI-AT'IC. [Gr.] Pertaining to the loins. The name of the large nerve of the loins and leg.

SCLE-ROT'IC. [Gr. σκληρος, *sklēros*, hard.] A membrane of the eye.

SE-BA'CEOUS. [L. *sebum*, tallow.] Pertaining to fat; unctuous matter.

SE-CRE'TION. The act of producing from the blood substances different from the blood itself, as bile, saliva; the matter secreted, as mucus, bile, etc.

SE-CRE'TO-RY. Performing the office of secretion.

SEM-I-CIR'CU-LAR. Having the form of a half circle. The name of a part of the ear.

SEM-I-LU'NAR VALVES. [L. *semi*, half, and *luna*, the moon.] Name of the three festooned valves of the heart at the entrance of the great arteries.

SEP'TUM, -A. [L.] A membrane that divides two cavities from each other.

SE'ROUS. Thin; watery. Pertaining to serum.

SE'RUM. [L.] The thin, transparent part of blood.

SER-RA'TED. [L. *serra*, a saw.] Notched on the edge like a saw. Name of muscles.

SIG'MOID. [Gr.] Resembling the Greek ς, Sigma.

SI'NUS. [L., a bay.] A cavity, the interior of which is more expanded than the entrance.

SKEL'E-TON. [Gr. σκελλω, *skellō*, to dry.] The aggregate of the hard parts of the body; the bones.

SPHE'NOID. [Gr. σφην, *sphēn*, a wedge, and ειδος, *eidos*, likeness.] A bone at the base of the skull.

SPHINC'TER. [Gr. σφιγγω, *sphingō*, to restrict.] A muscle that contracts or shuts an orifice.

SPI'NAL CORD. A prolongation of the brain.

SPINE. A thorn. The vertebral column; back-bone.

SPLANCH-NOL'O-GY. [Gr. σπλαγχνον, *splanchnon*, the bowels, and λογος, *logos*, a discourse.] A description of the internal parts of the body.

SPLEEN. A very vascular organ situated in the abdomen and attached to the stomach; the milt.

STA'PES. The name of one of the small bones of the ear.

STER'NUM. The breast-bone.

Stig'ma-ta. The apertures in the bodies of insects communicating with the air-vessels.

Stom'ach. The principal organ of the digestive apparatus.

Stra'tum, -a. [L. *sterno*, to spread.] A bed; a layer.

Sub-cla'vi-an. [L. *sub*, under, and *clavis*, a key.] Situated under the clavicle.

Sub-cu-ta'ne-ous. [L. *sub*, under, and *cutis*, skin.] Situated under the skin.

Sub-lin'gual. [L. *sub*, under, and *lingua*, the tongue.] Situated under the tongue.

Sub-max'il-la-ry. [L. *sub*, under, and *maxilla*, the jaw-bone.] Located under the jaw.

Su-pi-na'tor. [L.] A muscle that turns the palm of the hand upward.

Su'ture. [L. *suo*, to sew.] The seam or joint that unites the bones of the skull.

Syn-o'vi-a. [Gr. συν, *sūn*, with, and ωον, *ōn*, an egg.] The fluid secreted into the cavities or joints for the purpose of lubricating them.

Syn-o'vi-al. Pertaining to synovia.

Sys'tem. An assemblage of organs composed of the same tissues and intended for the same functions.

Sys-tem'ic. Belonging to the general system.

Tac'tile. [L. *tango*, I touch.] That may be felt; connected with the sense of touch.

Tar'sus. [L.] The posterior part of the foot. The instep.

Tem'po-ral. [L. *tempus*, time.] Pertaining to the temples; so called because the hair early begins to turn white with age in that portion of the scalp.

Ten'don. [Gr. τεινω, *teinō*, to stretch.] A fibrous cord by which a muscle is attached to a bone.

Ten-to'ri-um. [L. *tendo*, to stretch.] A process of the dura mater which lies between the cerebrum and cerebellum.

Tho'rax. [Gr.] That part of the skeleton that composes the bones of the chest; the cavity of the chest.

Tho-rac'ic. Relating to the chest.

Thy'roid. [Gr. θυρεος, *thureos*, a shield.] Resembling a shield. A cartilage of the larynx.

Tib'i-a. [L., a flute.] The large bone of the leg.

Tis'sue. A web-like structure constituting the elementary structure of animals.

Ton'sil. [L.] A glandular body in the throat or fauces.

Tra'chea, -æ. [Gr. τραχυς, *trachus*, rough.] The windpipe. In insects the air-tubes which ramify in the body.

Trans-verse'. Lying in a cross direction.

Tri'ceps. [L. *tres*, three, and *caput*, head.] Three. A name given to muscles that have three attachments at one extremity.

Tri-cus'pid. [1. *tres*, three, and *cuspis*, point.] The triangular valves in the right side of the heart.

Trit'u-rat-ing. Grinding to a powder.

Troch'le-a. [Gr. τροχαλια, *trochalia*, a pulley.] A pulley-like cartilage over which the tendon of a muscle of the eye passes.

Trunk. The principal part of the body to which the limbs are articulated.

Tu'ber-cle. [L. *tuber*, a bunch.] A pimple, swelling or tumor on animal bodies.

Tu'bule. [L.] A little tube or pipe.

Tur'bi-na-ted. [L. *turbo*, a whirling.] Three rolled or convoluted plates of bone attached to the outer wall of each nostril.

Tym'pan-um. [L.] The middle ear.

Ul'na. [It.] A bone of the forearm.

U-re'ter. [Gr. ουρειν, *ourein*, to conduct water.] The excretory duct of the kidneys.

Valve. Any membrane, or doubling of any membrane, which prevents fluids from flowing back in the vessels and canals of the animal body.

Vas'cu-lar. [L. *vasculum*, a vessel.] Pertaining to vessels; abounding in vessels.

Veins. Vessels that convey blood *to* the heart.

Ven'tral. [L. *venter*, the stomach.] Relating to the lower or inferior surface of the body.

Ven'tri-cle. [L. *venter*, the stomach.] A small cavity of the animal body.

Verm-i-form'is. [L. *vermis*, a worm, and *forma*, form.] Having the form and shape of a worm.

Vert'e-bra, -æ. [L. *verto*, to turn.] A joint of the spinal column.

Vert'e-brates. [L.] Animals which have an internal skeleton supported by a vertebral or spinal column.

Ves'i-cle. [L. *vesica*, a bladder.] A little bladder.

Ves'ti-bule. [L.] A porch of a house. A cavity belonging to the ear.

Vil'li. [L. *villus*, hair.] The thread-like projections from the inner surface of the membrane that lines the small intestines.

Vi'rus. [L., poison.] Foul matter of an ulcer; poison.

Vi'tal. [L. *vita*, life.] Pertaining to life.

Vit're-ous. [L. *vitrum*, glass.] Belonging to glass. A humor of the eye.

Vo'mer. [L., a ploughshare.] One of the bones of the nose.

Zo-ol'o-gy. [Gr. ζοών, *zoön*, an animal, λογος, *logos*, a discourse.] That branch of Natural History that treats of animals.

Zyg-o-mat'i-cus. [Gr. ζυγος, *zugos*, a yoke.] A term applied to some muscles of the face from their attachment.

INDEX.

ABSORBENTS, Anatomy of, 59, 75.
 Physiology of, 78.
 Hygiene of, 80.
ABSORPTION, 75.
AIR, 43, 91, 114.
AIR-CELLS, 84, 87.
ALVEOLAR PROCESSES, 51.
AMPHIBIANS, Definition of, 27.
ANATOMY, Definition of, 8.
ANGIOLOGY, Comparative, 124.
ANNULOSA, 25, 33, 72, 100, 127, 150.
AORTA, 116.
APPARATUS, Definition of, 8.
ARTERY Pulmonary, 85, 119.
ARTERY, 116.
ASPHYXIA, from Drowning, 180.
 from Carbonic Acid Gas, 180.
ASSIMILATION, 57, 130.
AURICLE, of the Heart, 115.

BATHING, 112, 171.
BEDS, Making of, 172.
BILE, 58.
BIRDS, Definition of, 26.
BLOOD, Temperature of, 89, 116.
BLOOD-VESSELS, 85.
BONES, Anatomy of, 13.
 Physiology of, 20.
 Hygiene of, 22.
BRAIN, 135.
BRONCHI, 84.
BURNS AND SCALDS, 178.

CAPILLARIES, 85, 119.
CARBONIC ACID, 87, 91.
CARPUS, 17.
CARTILAGE, 18.
CELLS, Definition of, 9.
CEREBELLUM, 135, 139.
CEREBRO-SPINAL SYSTEM, 134.
CEREBRUM, 135, 139.
CHEST, Compression of, 23, 95.

CHILBLAINS, 180.
CHYLE, 59.
CHYME, 59.
CIRCULATORY ORGANS, 115.
 Physiology of, 121.
 Hygiene of, 122.
CLAVICLE, 16.
CLOTHING, 109, 124.
COCCYX, 15.
COCHLEA, 164.
COMMISSURES of the Brain, 134.
CORIUM, 104, 108.
CORNS, Treatment of, 179.
CRYSTALLINE LENS, 157
CUTICLE, 103.

DERMIS, 104.
DIAPHRAGM, 85.
DIET, 144.
DIGESTIVE ORGANS, Anatomy of, 51.
 Physiology of, 57.
 Hygiene of, 59.
DROWNED PERSONS, Treatment of, 180.

EAR, 162, 164, 166.
EPIDERMIS, 103, 107.
EPIGLOTTIS, 82.
EPITHELIAL CELLS, 9, 57.
EXCRETION, 131.
EYE, 156, 159, 162.
 Protecting Organs of, 158.

FACE, Bones of, 14.
FASCIA, 37.
FASCICULI, 36.
FELON, 24.
FEMUR, 15.
FIBULA, 15.
FIBRE, 36.
FISHES, Definition of, 27.
FOOD, Quantity of, 60.
 Quality of, 61.

INDEX.

FOOD, Manner of taking, 62.
    Conditions when received, 63.
FROST-BITE, Treatment of, 179.
FUNCTION, 8.
FURCULA, 30.

GANGLIONS, 133.
GASTRIC JUICE, 58.
GLANDS, Definition of, 52.
    Lachrymal, 159.
    Lymphatic, 77.
    Mesenteric, 77.
    Oil, 105, 108.
    Perspiratory, 106.
    Salivary, 53.
    Secreting, 131.

HAIR-FOLLICLES, 105, 109.
HEART, 115.
    Auricles of, 116.
    Ventricles of, 116.
HEAT, Animal, 89, 132.
HEMORRHAGE, Means of Arresting, 177.
HUMERUS, 16.
HYDROGEN, 89.
HYGIENE, Definition of, 8.

INSECTS, 127.
INTESTINES, 54.
INTESTINAL JUICE, 58.
INVERTEBRATES, 25, 40, 71, 126, 150.
IRIS, 157.

JOINTS, 17, 21.

KIDNEYS, 131.

LABYRINTH, 163.
LACHRYMAL APPARATUS, 159.
LACTEALS, 77.
LARYNX, 82, 90.
LIGAMENT, 18.
LIGHT, 43, 114.
LIVER, 55.
LUNGS, 84.
LYMPH, 78.
LYMPHATICS, 75, 108.
    Duct, 77.

MAMMALS, Definition of, 26.
MEDICINE, 173.
MEDULLA, 19.
    Oblongata, 134.
MEMBRANE, Basement, 9.
    Of the Brain, 135.

MEMBRANE, Mucous, 9.
    Secreting, 131.
    Serous, 9.
    Synovial, 9, 16.
METACARPUS, 17.
METATARSUS, 16.
MOLLUSCA, 25, 33, 72, 101, 127 151.
MOTORY APPARATUS, 12.
MOUTH, Structure of, 51.
MUCUS, 9, 58.
MUSCLES, Anatomy of, 36.
    Physiology of, 39.
    Hygiene of, 42.
    Of eye, 158.
    Respiratory, 85.
    Intercostal, 86.
MYOLOGY, Comparative, 48.

NAILS, 106, 109.
NERVE CENTRES, 139.
NERVES, 133, 136.
    Cranial, 136.
    Cutaneous, 105.
    Spinal, 137.
    Sympathetic, 138, 140.
NERVOUS SYSTEM, Anatomy of, 133.
    Physiology of, 139.
    Hygiene of, 142.
NEURILEMA, 136.
NEUROLOGY, Comparative, 148.
NITROGEN, 87.
NURSE, Duty of, 171.
NUTRITIVE APPARATUS. 51.

ŒSOPHAGUS, 53.
OIL-GLANDS, 105.
ORGAN, Definition of, 8.
OSTEOLOGY, Comparative 24.
OXYGEN, 87.

PAPILLÆ, 104 108.
PALATE, 51.
PANCREAS, 55.
PANCREATIC JUICE, 58.
PATELLA, 15.
PELVIS, 14.
PERICARDIUM, 115.
PERICRANIUM, 19.
PERIOSTEUM, 19.
PERSPIRATION, Uses of, 100.
PHALANGES, 16, 17.
PHARYNX, 83.
PHYSIOLOGY, Definition of, 8.
PLEURA, 84.

INDEX.

PLEXUS, 139.
PNEUMONOLOGY, Comparative, 97.
POISONS and their Antidotes, 180.
PROTOZOA, 25, 34, 74, 101, 128.

RADIATA, 25, 34, 74, 128, 151.
RADIUS, 16.
REPTILES, Definition of, 27.
RESPIRATION, Artificial, 180.
RESPIRATORY ORGANS, Anatomy of, 82.
    Physiology of, 86.
    Hygiene of, 91.
RETINA, 157.
RIBS, 14.

SACRUM, 15.
SALIVA, 58.
SCAPULA, 16.
SECRETION, 131.
SICK-ROOM, 93, 172.
SINUSES, Venous, 127.
SKELETON, 18.
SKIN, Anatomy of, 103.
    Physiology of, 107.
    Hygiene of, 109.
SKULL, 13.
SLEEP, 146.
SLEEPING-ROOM, 93.
SOUND, 165.
SMELL, Sense of, 154, 155.
SPINAL CORD, 134, 140.
    " COLUMN, 14.
SPLANCHNOLOGY, Comparative, 95.
SPLEEN, 55.
STERNUM, 14.
STIGMATA, 100.
STOMACH, 54.
SWEAT GLANDS, 106.
SYNOVIA, 9.

TARSUS, 16.
TASTE, Sense of, 167.
TEETH, 51.
TENDONS, 37, 40.
THORACIC DUCT, 75.
TIBIA, 15.
TISSUES, Adipose, 19.
    Cellular, 9.
    Muscular, 9.
    Nervous, 9.
TONGUE, 167.
TOUCH, Sense of, 168.
TRACHEA, 83.
TRUNK, 14.
TYMPANUM, 162.

ULNA, 16.
URETER, 131.

VALVES of the Heart, 116.
VEINS, 119.
    Absorbent, 78.
    Hepatic, 121.
    Portal, 78, 121.
    Pulmonary, 85, 120.
VENTILATION, 92, 173.
VENTRICLES of the Heart, 116.
    Of the larynx, 83.
VERTEBRA, 14.
VERTEBRATES, 25, 26, 48, 66, 97, 121, 142, 154, 167, 169.
VESTIBULE, 164.
VILLI, 56, 59.
VOCAL CORDS, 83, 90.
    " ORGANS, 82, 91, 96.

WATCHER, Duty of, 175.
WONDER-NETS, 125.
WOUNDS, Treatment of, 176.

KEY TO CUTTER'S

NEW OUTLINE ZOOLOGICAL CHARTS,

OR

HUMAN AND COMPARATIVE ANATOMICAL PLATES.

SUGGESTIONS TO TEACHERS.

In using these charts, we would suggest that the pupil carefully examine the illustrating cuts interspersed with the text in connection with the lesson to be recited. The similarity between these and the charts will enable the pupil to recite, and the teacher to conduct his recitation from the latter.

Let a pupil show the situation of an organ, or part, on an anatomical outline chart, and also give its structure, while other members of the class note all omissions and misstatements. Another pupil may give the use of that organ, and, if necessary, others may give an extended explanation. The third may explain the laws on which the health of the part depends, while other members of the class may supply what has been omitted. After thus presenting the subject in the form of topics, questions may be proposed promiscuously from each paragraph, and where examples occur in the text let other analogous ones be given.

If the physiology and hygiene of a given subject have not been studied, confine the recitation to those parts only on which the pupil is prepared. When practicable, the three departments should be united; but this can only be done when the chapter on the hygiene has been learned, while the physiology can be united with the anatomy in all chapters upon physiology.

CHART No. 1.

OSSEOUS SYSTEM—HUMAN AND COMPARATIVE.

A. *Bones of the Human Body.*—1, The frontal bone. 2, The superior maxillary (upper jaw-bone). 3, The inferior maxillary (lower jaw-bone). 4, The cervical vertebræ (bones of the neck). 5, 6, The lumbar vertebræ (bones of the loins). 7, The sacrum. 8, The temporal bone. 9, The scapula (shoulder-blade). 10, 10, 10, The ribs. 11, 11, The innominata (hip-bones). 12, The humerus (arm-bone). 13, The radius. 14, The ulna. 15, The carpus (wrist-bones). 16, 16, The metacarpus (bones of the palm of the hand). 17, 17, The phalanges (finger-bones). 18, The femur (thigh-bone). 19, The patella (knee-pan). 20, The tibia. 21, The fibula. 22, The tarsus (bones of the instep). 23, 23, The metatarsus (bones of the middle of the foot). 24, 24, The phalanges (toe-bones). 25, Ligaments of the shoulder. 26, Ligaments of the elbow. 27, Ligaments of the wrist. 28, Ligaments of the hip-joint. 29, Ligaments of the knee. 30, Interosseous membrane. 31, Ligaments of the ankle. 32, The clavicle (collar-bone). 33, The sternum (breast-bone).

B. *Bones of the Cow.*—1, The frontal bone. 2, The upper jaw (superior maxillary). 3, The lower jaw (inferior maxillary). 4, The cervical vertebræ (bones of the neck). 5, The dorsal vertebræ (bones of the back). 6, 7, The lumbar vertebræ. 8, The caudal vertebræ. 9, The scapula. 10, 10, The ribs. 11, The innominate bones. 12, The humerus. 13, 14, The radius and ulna. 15, The carpus. 16, The metacarpus. 17, The phalanges. 18, The femur. 20, The tibia. 22, The tarsus. 23, The metatarsus. 24, The phalanges.

C. *Bones of the Bird.*—1, The cranium. 2, The superior mandible (upper jaw). 3, The interior mandible (lower jaw). 4, The cervical vertebræ. 5, The dorsal vertebræ. 8, The coccyx. 9, The scapula. 10, The ribs. 11, The pelvis. 12, The humerus. 13, 14, The radius and ulna. 15, The carpus. 16, The metacarpus. 17, 17, Phalanges. 18, The femur. 20, The tibia. 21, The fibula. 22, 23, The metatarsus. 24, Phalanges. 32, The coracoid bone. 33, The clavicle (furcula). 34, The sternum.

D. *Bones of the Tortoise.*—1, The head. 4, The cervical vertebræ. 5, 5, 6, 6, The dorsal vertebræ and lateral plates. 7, The iliac bones. 8, The caudal vertebræ. 9, The scapula. 12, The humerus. 13, 14, The radius and ulna. 15, The carpus. 16, 17, Phalanges. 18, The femur. 20, The tibia. 21, The fibula. 22, The tarsus. 23, 24, Phalanges. 32, The clavicle. 33, The coracoid bone.

E. *Bones of the Fish.*—1, The bones of the head. 2, The upper jaw. 3, The lower jaw. 4, 5, 6, The dorsal and caudal vertebræ. 8, The first dorsal fin. 9, The second dorsal fin. 10, One of the ventral fins. 12, A pectoral fin. 18, A ventral fin.

F. *Diagram of an Annulose.*—1, The vascular (blood-vessel) system. 2, The digestive system. 3, 3, The ganglia (nervous) system. 4, 4, A series of rings of hardened skin which forms an external skeleton.

G. *Diagram of a Mollusk.*—1, The digestive canal. 2, The heart. 3, 4, 5, Ganglia (knots of nervous matter).

H. *Diagram of a Radiate.*—(A star-fish). 1, Central aperture.

CHART No. 2.

MUSCULAR SYSTEM—HUMAN AND COMPARATIVE.

A. *Muscles of Human Body.*—1, The occipito-frontalis. 2, The orbicularis palpebrarum. 3, The levator labii superioris. 4, The zygomaticus. 5, The masseter. 6, The orbicularis oris. 7, The temporal. 8, Zygomatici. 9, The depressor labii inferioris. 10, The deltoid. 11, 11, The pectoralis major. 13, The supinator longus. 14, Palmaris longus. 15, The flexor carpi radialis. 16, The obliquus externus. 17, The sartorius. 18, The adductor longus. 19, The rectus femoris. 20, The vastus externus. 21, The vastus internus. 22, The tendon of the quadriceps extensor. 23, The gastrocnemius. 24, The extensor longus digitorium. 25, The tibialis anticus. 26, The short extensor muscles of the toes. 27, The tendons of the long extensors. 28, The serratus magnus. 29, 29, The external abdominal rings. 30, The saphenous opening. 31, 31, 31, 31, The tendons of the wrist and fingers. 32, The sterno-hyoideus. 33, The sterno-cleido-mastoideus. 34, The biceps. 35, The triceps muscle.

B. *Muscles of the Cow.*—1, The occipito-frontalis. 2, The orbicularis palpebrarum. 3, The masseter. 4, The levator labii inferioris. 5, The platysma. 7, The trapezius. 10, The latissimus dorsi. 11, The pectoralis. 16, 17, The external and internal oblique muscle. 18, The opening for the mammary artery and vein (milk-veins). 19, The biceps femoris. 20, 20, 20, The gluteii muscles. 33, The masto-humeralis.

C. *Muscles of the Bird.*—1, The occipito-frontalis. 2, The orbicularis palpebrarum. 5, The masseter. 7, The temporal. 10, The deltoid. 11, The pectoralis. 13, The sacro-lumbalis. 14, The extensor carpi ulnaris. 19, The vastus externus. 20, The gluteii. 23, The flexor longus digitorium. 24, The extensor longus digitorium. 33, The longus colli. 34, The extensor plicæ alaris. 35, The teres major.

D. *Muscles of the Tortoise.*—1, The digastricus. 10, 10, The deltoides. 14, The palmaris. 18, The semi-membranosus. 23, The tibialis anticus. 24. The gastrocnemius. 28, The sub-coracoideus. 31, 32, The flexores digitorium. 34, 35, The triceps brachii.

E. *Muscles of the Fish.*—1, 2, 3, and a, b, c, represent the zigzag arrangement of the muscles of the fish (myocomma).

F. *Diagram of an Insect.*—1, The head. 2, The first segment of the chest, with the first pair of legs. 3, The second segment, with the second pair of legs and the first pair of wings. 4, The third segment, with the third pair of legs and second pair of wings. 5, The abdomen without legs.

CHART No. 3.
NUTRITIVE SYSTEM—HUMAN AND COMPARATIVE.

A. *The Internal Organs of Man.*—1, The parotid gland. 2, The submaxillary gland. 3, The sublingual gland. 4, The œsophagus. 5, The larynx and trachea. 6, The left lung. 7, The right lung. 8, The heart. 9, The vena cava descendens. 10, The aorta. 11, The pulmonary artery. 12, The stomach. 13, 14, The left and right lobe of the liver. 15, 15, 15, The large intestine. 16, 16, 16, 16, The small intestine. 17, The diaphragm. 18, The gall-bladder.

B. *Internal Organs of a Goat.*—1, The second stomach (reticulum). 2, The third stomach. 3, The fourth stomach (rennet). 4, Fold of the mesentery. 5, The jejunum. 6, The ileum. 7, The cæcum. 8, The colon. 9, The right kidney. 10, The rectum. 11, 12, Lobes of the liver (turned forward). 13, The gall-cyst. 14, Inferior part of abdomen. 15, The omentum.

C. *Organs of a winged Reptile.*—1, The ventricle of the heart. 2, 3, The auricles of the heart. 4, 5, 6, Blood-vessels. 7, The trachea. 9, 10, 11, The liver and its appendages. 12, The stomach. 13, The duodenum. 14, 15, 16, The intestines. 17, The cloaca. 18, The cæca.

D. *Diagram of the Organs of a Frog.*—1, The heart. 2, 2, Arches of the aorta. 3, 3, Pulmonary artery. 4, 4, The pulmonary veins. 6, The stomach. 5, The digestive canal.

CHART No. 4.
DIGESTIVE SYSTEM—HUMAN AND COMPARATIVE.

A. *Digestive Organs of Man.*—1, The upper jaw. 2, The lower jaw. 3, The tongue. 4, The hard palate (roof of the mouth). 5, The parotid gland. 6, The sublingual gland. 7, The larynx. 8, 9, The œsophagus. 10, The stomach. 11, 11, The liver. 12, The gall-bladder. 13, Its duct. 14, The duodenum. 15, The pancreas. 16, The spleen. 17, 17, 17, 17, The small intestine. 18, The cæcum. 19, The appendix vermiformis. 20, 20, The ascending colon. 21, The transverse colon. 22, 22, The descending colon. 23, The sigmoid flexure of the colon. 24, The rectum.

B. *Digestive Organs of a Fowl.*—9, The œsophagus. 8, The crop (ingluvies). 7, The second stomach (proventriculus). 10, The gizzard. 11, 11, The liver. 12, The gall-bladder. 13, The bile ducts. 14, 14, 14, 14, The duodenum. 15, The pancreas. 16, The cæca (pouches). 17, The large intestine. 24, The cloaca. 25, The trachea.

C. *Digestive Organs of an Ox.*—1, The œsophagus. 2, 2, The rumen (paunch). 3, The second stomach (reticulum). 4, The omasum (maniplies). 5, The fourth stomach or abomasum (rennet). 6, The duodenum (intestine).

D. *Digestive Organs of an Insect.*—8, The crop. 9, The gullet. 10, The gizzard. 14, 14, The chylific (digestive) stomach. 16, 16, Biliary vessels. 17, The intestine. 18, The renal vessels. 24, The cloaca.

E. *Digestive Organs of the Sword-Fish.*—11, 11, The liver. 13, The bile duct. 16, 16, The cæcas (pouches). 17, 17, 17, The intestine. 24, The large intestine.

F. *Digestive Organs of the Herring.*—1, 1, The air-bladder. 2, The air-duct (pneumatic). 9, The œsophagus. 10, The stomach. 16, The cæca. 17, 17, 17, The intestine.

CHART No. 5.
ABSORPTIVE SYSTEM—HUMAN AND COMPARATIVE.

A. *Absorbent Vessels in Man.*—1, 2, 3, 4, Lymphatic vessels and glands of the lower extremities. 5, 6, Inguinal lymphatics and glands. 8, Lymphatic vessels of the kidney. 12, The thoracic duct. 10, 10, 10, The intercostal lymphatics. 11, The receptaculum chyli. 13, Lymphatics of the neck. 14, 14, Carotid arteries. 15, Axillary glands. 16, 17, 18, Lymphatics of the arm and hand. 19, Lymphatics of the face. 20, The right subclavian vein. 21, The junction of the thoracic duct with the left subclavian vein.

B. *Section of the Layers of the Skin.*—1, The dermis. 2, 3, The epidermis. 4, The rete mucosum. 5, Subcutaneous connective and adipose tissue. 6, Tactile papillæ. 7, Sweat or perspiratory glands. 8, The duct of the sweat glands. 9, Spiral passages of the ducts through the epidermis. 10, 10, The termination of the ducts on the surface of the epidermis.

C. *Section of the Papillæ and Glands of the Skin.*—1, 1, 1, 1, Ridges of the cuticle (cut vertical). 2, 2, 2, Furrows or wrinkles of the cuticle. 3, The epidermis. 4, The rete mucosum. 5, The dermis. 6, 6, 6, The papillæ. 7, 7, Small furrows between the papillæ. 8, 8, 8, 8, Deeper furrows between each couple of the papillæ. 9, Fat cells. 10, 10, 10, The adipose layer, with numerous fat vesicles. 11, 11, Cellular fibres of the adipose tissue. 12, Two hairs. 13, Sweat or perspiratory gland, with its spiral duct. 14, A sudoriferous gland with a duct less spiral. 15, 15, Oil-glands, with ducts opening into the sheath of the hair.

CHART No. 6.

RESPIRATORY SYSTEM—HUMAN AND COMPARATIVE.

A. *Respiratory Organs of Man.*—1, The larynx. 2. The trachea. 3, The right bronchia. 4, The left bronchia. 5, 6, 7, Lobes of the right lung. 8, 9, Subdivisions of the bronchi or bronchial tubes. 10, 10, 10, 10, Air cells. 11, 11, The diaphragm.

B. *Diagram of the Blood-vessels in Man.*—1, The vena cava descendens. 2, The vena cava ascendens. 3, The right ventricle of the heart. 4, The left ventricle. 5, 6, The aorta. 7, The pulmonary artery. 8, 9, Divisions of the pulmonary artery. 11, Pulmonary vein.

C. *Section of a Quadruped.*—1, The œsophagus. 2, The trachea. 5, 6, The lungs. 7, The heart. 8, The stomach. 9, The liver. 10, 10, Intestines. 11, 11, The diaphragm. 12, 13, The kidney and duct. 14, The brain. 15, 15, 15, The spinal cord. 16, 16, 16, The vertebræ.

D. *Section of a Lobule of a Bird's Lung.*—2, A bronchial tube. 3, 4, Divisions of a bronchus that end in sacs. 8, 8, 9, 9, Abdominal air-sacs.

E. *Lung of a Goose.*—2, A bronchus. 3, 4, The bronchial tubes laid open. 10, 10, Apertures of communication with air-cells. 11, 11, Abdominal bronchial orifices.

G. *Respiratory Organs of the Water-scorpion.*—1, The head. 2, The base of the first pair of feet. 3, The first ring of the thorax. 4, The base of wings. 5, Base of the second pair of feet. 6, 6, 6, 6, Stigmata (opening at the edge of each joint). 7, 7, 7, 7, Tracheæ (air-tubes). 8, 8, Air-sacs.

F. *Diagram of the Bronchial Leaflets of the Cod.*—1, A section of a bronchial arch. 2, 3, Bronchial leaflets or plates.

J. *Diagram of the Circulation of the Blood through the Bronchial Leaflets.*—1, A section of a bronchial arch. 2, A section of a bronchial artery. 3, 3, An arterial branch along the outer margin of the processes, giving off capillary vessels to the leaflets. 4, A vein that receives the blood from the capillaries of the inner margin of the process. 5, Bronchial vein.

H. *A Plexus of Capillary Vessels.*

K. *Diagram of the Relative Positions of the Blood-vessels to the Air-cells.*— 1, A bronchial tube communicating with the air-cells, 2, 2, 2. 3, A branch of the pulmonary artery containing bluish blood. 4, A branch of a pulmonary vein containing scarlet or purified blood.

CHART No. 7.

CIRCULATORY SYSTEM—HUMAN AND COMPARATIVE.

A. *Circulation in Man.*—3, The right ventricle. 4, The right auricle. 5, Arch of aorta. 6, Left pulmonary artery. 7, The vena cava descendens. 8, The vena cava ascendens. 9, The descending aorta. 10, The right femoral artery. 11, The left femoral vein. 12, The subclavian artery. 13, The subclavian vein. 14, The jugular vein. 15, The basilic vein. 16, The cephalic vein. 17, The kidney. 18, The brachial artery. 19, The ulnar artery. 20, The radial artery. 21, The anterior tibial artery. 22, The posterior tibial artery.

B. *Diagram of the Circulation in Reptiles.*—1, Ventricle. 2, 3, Left auricle. 4, Right auricle. The arrows show the direction of the blood.

C. *Diagram of the Circulation in the Fish.*—1, The pericardium. 2, The auricle that receives blood from the body. 3, The ventricle that sends blood to the gills.

D. *Diagram of the Heart of Mammals.*—1, The vena cava descendens. 2, The vena cava ascendens. 3, The right auricle. 4, The opening between the right auricle and right ventricle. 5, The right ventricle. 6, The tricuspid valve. 7, The pulmonary artery. 8, 8, Its branches. 9, The semi-lunar valves of pulmonary artery. 10, The septum between the two ventricles of the heart. 11, 11, The pulmonary veins. 12, The left auricle. 13, The opening between the left auricle and the left ventricle. 14, The left ventricle. 15, The mitral valve. 16, The aorta. 17, The semi-lunar valves of the aorta.

E. *The Heart and Arteries of a Snail.*—2, The stomach. 3, 3, The intestine. 5, The heart. 6, The aorta. 7, The pulmonary artery.

CHART No. 8.

NERVOUS SYSTEM—HUMAN AND COMPARATIVE.

A. *Section of the Human Brain and Spinal Column.*—1, The cerebrum. 2, The cerebellum. 3, The medulla oblongata. 4, 4, The medulla spinalis (spinal cord) in the canal formed by the vertebræ of the spinal column.

B. *Back view of the Brain and Nerves in Man.*—1, The cerebrum. 2, The cerebellum. 3, The spinal cord. 4, Nerves of the face. 5, Brachial plexus of nerves. 6, Internal cutaneous. 7, Ulnar. 8, Musculo-spiral. 9, Circumflex. 10, Intercostal. 11, Lumbar plexus. 12, Sacral plexus. 13, Posterior tibial. 14, Anterior tibial. 15, Popliteal. 16, Sciatic. 17, Coccygeal.

C. *The Sympathetic Nerves.*—1, The renal plexus of nerves. 2, 3, 4, Lumbar ganglion. 5, Aortic plexus. 6, Solar plexus. 7, Dorsal ganglia. 8, 9,

Cardiac nerves. 10, Inferior cervical ganglia. 11, Brachial plexus. 12, Superior cervical ganglia.

D. *Base of the Brain of a Horse.*—1, The cerebrum. 2, The optic ganglion. 3, The cerebellum. 4, The medulla oblongata and spinal cord.

E. *Brain of an Alligator.*—1, The olfactory ganglion. 2, The cerebrum. 3, The optic ganglion. 4, The cerebellum. 5, The medulla oblongata and spinal cord.

F. *Brain of a Bird.*—1, The cerebrum. 2, The optic ganglion. 3, The cerebellum. 4, The medulla oblongata.

G. *Brain of a Fish.*—1, The olfactory ganglion. 2, The cerebrum. 3, The optic ganglion. 4, The cerebellum. 5, The medulla oblongata and spinal cord.

H. *Nervous System of the Beetle.*—1, 1, 2, 2, Nervous ganglions and cords.

I. *Diagram of the Nervous System of the Centipede.*—1, Nervous ganglia.

J. *Diagram of the Nervous System of the Star-Fish.*

CHART No. 9.
SPECIAL SENSE—HUMAN AND COMPARATIVE.

A. *The Nervous System of Man.*—The convolutions of the large brain (cerebrum). 2, The lesser brain (cerebellum). 3, The cervical nerves. 4, The dorsal nerves. 5, The lumbar nerves. 6, The sciatic. 7, The peroneal nerve. 8, The posterior tibial nerve. 9, Median nerve.

B. *Section of the Globe of the Eye.*—1, The choroid coat of the eye. 2, The sclerotic coat. 3, The retina. 4, The cornea. 5, 5, The iris. 6, The pupil. 8, 9, The chambers of the eye that contain the aqueous humor. 10, The crystalline lens. 11, 11, The vitreous humor. 12, Arteria centralis retinæ. 13, The optic nerve.

C. *Distribution of the Trifacial (fifth pair) Nerve.*—1, The trifacial nerve. 2, A branch that passes to the eye (ophthalmic). 3, A branch distributed to the teeth of the upper jaw (superior maxillary). 4, The branch that passes to the tongue (5) and teeth of the lower jaw (the inferior maxillary). 6, The gustatory branch. 7, Inferior dental nerve.

D. *Distribution of the Olfactory Nerve.*—1, The olfactory (or nerve of smell). 2, 2, The fine divisions of this nerve on the membrane of the nose. 3, A branch of the fifth pair (trifacial) nerve.

E. *Front view of the Organ of Hearing.*—1, The auditory canal. 2, The drum of the ear (membrana tympani). The chain of bones in the ear (3, The malleus. 4, The incus, and, 5, The stapes). 6, The cavity of the tympanum. 7, The vestibule. 8, 9, 10, The semi-circular canals. 11, 11, 12, Channels of the cochlea. 13, Cavity in the mastoid portion of temporal bone. 14, The opening from the middle ear to the throat (Eustachian tube).

F. *Compound Eye of the Bee.*— Its division into facets (highly magnified).

F. Facets still more highly magnified.

F. Facets with hairs growing between them.

VALUABLE EDUCATIONAL WORKS
FOR
SCHOOLS, ACADEMIES, AND COLLEGES.

Selected from Messrs. J. B. LIPPINCOTT & Co.'s Catalogue, which comprises nearly Two Thousand Works in all branches of Literature. Catalogues furnished on application. *Liberal terms will be made for Introduction.*

Haldeman's Outlines of Etymology. By S. S.
Haldeman, A.M., author of "Analytical Orthography," "Elements of Latin Pronunciation," etc. 12mo. Fine cloth. 90 cents.

"This is a most scholarly presentation of the science of etymology, . . . to which is added an appendix of inestimable value to the student of language. . . . It is a marvel of conciseness."—*New England Journal of Education*, July 5, 1877.
"There is, probably, no man living who has studied and analyzed the English language so thoroughly and so successfully as this distinguished savant of Pennsylvania, and this little treatise is the latest fruit of his ripe scholarship and patient research. No newspaper notice can do justice to the work, for it cannot be described, and must be studied to be appreciated."—*Philadelphia Evening Bulletin.*

Walker's Science of Wealth. A Manual of
Political Economy, embracing the Laws of Trade, Currency, and Finance. Condensed and Arranged expressly for Use as a Text-book. By AMASA WALKER, LL.D., *late Lecturer on Public Economy, Amherst College.* Student's edition. 12mo. Extra cloth. $1.50.

"I have, during the past year, made use of Dr. A. Walker's 'Science of Wealth' in the new and condensed form he has given it, in giving instruction to the senior class in Political Economy, and have found the book better adapted than any other with which I am acquainted for use in a college class-room.
It is clear, compact, and ample in its illustrations."—PROF. JULES H. SEELYE, *Amherst College, Massachusetts.*
"It is, in my opinion, the best work on this subject."—W. T. HARRIS, *Superintendent of Schools, St. Louis, Missouri.*

Berkeley's Principles of Human Knowledge. A
Treatise concerning the Principles of Human Knowledge. By GEORGE BERKELEY, D.D., *formerly Bishop of Cloyne.* With Prolegomena, and with Illustrations and Annotations, Select, Translated, and Original. By CHARLES P. KRAUTH, D.D., *Norton Professor of Systematic Theology and Church Polity in the Evangelical Lutheran Theological Seminary; Professor of Intellectual and Moral Philosophy, and Vice-Provost of the University of Pennsylvania.* 8vo. Extra cloth. $3.00.

Prof. A. C. Fraser, of the University of Edinburgh, says of "Berkeley's Principles:" "It would be difficult to name a book in ancient or modern philosophy which contains more fervid and ingenious reasoning than is here employed to meet supposed objections, or to unfold possible applications to religion and science."

Neely's Elementary Speller and Reader. Containing the Principles and Practice of English Orthography and Orthoepy systematically developed. Designed to accord with the "present usage of literary and well-bred society." In three parts. By REV. JOHN NEELY. *Fourth Edition*, carefully revised. 16mo. Boards. 20 cents.

Turner on Punctuation. A Hand-Book of Punctuation, containing the more Important Rules, and an Exposition of the Principles upon which they depend. By JOSEPH A. TURNER, M.D. New, Revised Edition. 16mo. Limp cloth. 75 cents.

Smith's New Geography. Containing a concise Text and Explanatory Notes. Based on a Combination of the Analytical, Synthetical, and Comparative Systems. With more than 100 Maps of Religion, Government, Civilization, Races, Countries, Roman Empire, Vicinities, Rain, Wind, Seasons, Isothermals, Solar System, etc., and combining with much new and valuable matter many features not found in any other work of its class. By ROSWELL C. SMITH. Quarto. Half bound. $1.75.

"Simple and concise, but not dry; philosophical, yet practical. Combining more of the essential requisites of a Grammar and High School Geography than any work extant."

Allen's Comprehensive Geography. Combining Physical, Mathematical, and Political Geography with Important Historical Facts. Illustrated with numerous accurate Maps and Engravings. By F. A. ALLEN and B. F. SHAW. 4to. Boards. $1.60.

The early animals whose remains are numerous are objects of curiosity, the vegetation to which we trace the coal formations, the later plants and the higher animals, including man; the great empires of antiquity; the theories of the ancients concerning the earth; the results of modern investigation; the political divisions of the present day,—these are spoken of in their natural order. This chronological arrangement facilitates the elucidation of the mathematical part of the study. It enables the pupil to see the earth as the ancients saw it; to change his ideas as mankind changed theirs; and to regard the terrestrial mass as men regard it now. Instead of exhibiting the globe at the outset, it assists the reasoning powers in slowly forming into a round body the apparently flat expanse of land and water.

Allen's Primary Geography. On the basis of the Object Method of Instruction. By F. A. ALLEN, *Principal of Pennsylvania State Normal School, Mansfield, Pa.* Illustrated with Maps and Engravings. 4to. Boards. 60 cents.

Truthful pictures are the nearest representations of objects. By their aid I commenced with the pupil himself, surveyed his common surroundings, rambled and journeyed with him from place to place, from State to State, from continent to continent, viewing noteworthy objects, speaking of their uses, mentioning leading facts about seas, lakes, rivers, mountains, animals, and plants, in language to interest and in a manner to incite.

Johnson's Analytical Geometry. An Elementary Treatise, embracing Plane Co-ordinate Geometry, and an Introduction to Geometry of Three Dimensions. Designed as a Text-Book for Colleges and Scientific Schools. By WM. WOOLSEY JOHNSON, B.A., *Assistant Professor of Mathematics, U. S. Naval Academy.* 12mo. Cloth. $1.80.

"This is one of the simplest as well as most intelligible and practical books on Exact Science that has come under our notice."—*Philadelphia Press.*

"I am convinced that it is an excellent work, and well calculated for a text-book for colleges and scientific schools."—PROF. N. M. CRAWFORD, *Georgetown College, Ky.*

"It is superior to all text-books of the same class that have as yet come under my observation, in clearness of expression and well-chosen illustrations of general solutions."—C. HOMUNG, *Prof. of Mathematics in Hillsboro' College.*

"It is eminently suited to the wants of all students."—*College Courant.*

Hawes's Manual of United States Surveying. A System of Rectangular Surveying employed in Subdividing the Public Lands of the United States, etc. Illustrated with Forms, Diagrams, and Maps. Constituting a complete Text-Book of Government Surveying. By J. H. HAWES, *late Principal Clerk of Surveys in the General Land Office.* Crown 8vo. Extra cloth. $3.00.

"This book embodies in a complete form all the varied information so often sought after by county surveyors and others in regard to the system in use by the United States for surveyings, subdividing sections, running and making boundaries, etc."—*New Orleans Times.*

"This volume contains the system of rectangular surveying employed in subdividing sections and restoring lost corners of the public lands. The volume is compact and handsome, and will be found to answer its purpose admirably."—*Chicago Tribune.*

Bledsoe's Philosophy of Mathematics. With Reference to Geometry and the Infinitesimal Method. By A. T. BLEDSOE, LL.D., *late Professor of Mathematics in the University of Virginia.* 12mo. Cloth. $1.75.

"The author was professor of mathematics in the University of Virginia. This is an evidence of first-class abilities as a geometer. The object of the book is to shed clearer scientific light on the Infinitesimal Calculus."—*Universe, Philadelphia.*

"A new treatise on the calculus, in which the whole philosophy of the higher mathematical branches is followed out from the first appearance of its elements in Greek geometry."—*New Orleans Times.*

Hallowell's Geometrical Analysis; or, the Construction and solution of Various Geometrical Problems from Analysis, by Geometry, Algebra, and Differential Calculus; also, the Geometrical Construction of Algebraic Equations, and a Mode of Constructing Curves of the Higher Order by means of Points. With Portrait from steel. 8vo. Cloth extra. $2.50.

Playfair's Euclid. From the latest London Edition. Revised and Corrected. By J. PLAYFAIR, F.R.S., L. and E. 12mo. Half roan. $1.50.

Atwater's Elementary Logic. Designed Especially for the Use of Teachers and Learners. By L. H. ATWATER, *Professor of Mental and Moral Philosophy in the College of New Jersey.* 12mo. Cloth. $1.25.

"This manual shows the hand of a master, and will speedily take its proper place as a valuable contribution to the cause of liberal education."—*Congregational Quarterly.*

"I observe in Atwater's excellent Manual of Elementary Logic a disposition to unite the real improvements of the analytic with the established truths of the old logic."—DR. JAMES MCCOSH, *late Professor of Logic in Queen's College, Belfast, Ireland.*

Dr. Blair's Lectures on Rhetoric. Abridged, with Questions. 60 cents.

The want of a system of Rhetoric upon a concise plan has rendered this little volume acceptable to the teaching and reading public. Many who are terrified at the idea of travelling over a ponderous volume in search of information, will yet set out on a short journey in pursuit of science with alacrity and profit.

Samson's Art Criticism. Comprising a Treatise on the Principles of Man's Nature as Addressed by Art; together with a Historic Survey of the Methods of Art Execution in the Departments of Drawing, Sculpture, Architecture, Painting, Landscape Gardening, and the Decorative Arts. Designed as a Text-Book for Schools and Colleges, and as a Hand-Book for Amateurs and Artists. By G. W. SAMSON, *President of Columbia College.* 8vo. Cloth. $3.15. ABRIDGED EDITION. 12mo. Cloth. $1.60.

"Art criticism, boiled down into small doses for amateurs and students, is the function of Dr. G. W. Samson, President of Columbia College, Washington."—*Philadelphia Bulletin.*

"This work should be in the possession of every lover of the beautiful, and the abridgment deserves to be introduced into all our high schools, and made a part of common education."—*North American.*

Malcom's Butler's Analogy. The Analogy of Religion to the Constitution and Course of Nature. To which are added Two Brief Dissertations. I. On Personal Identity. II. On the Nature of Virtue. By JOSEPH BUTLER, D.C.L. With Introduction, Notes, Conspectus, and ample Index. Prepared by HOWARD MALCOM, D.D., LL.D., *President of the University at Lewisburg.* 12mo. Extra cloth. $1.15.

"We fully justify the editor's reasons for offering another edition of Butler, viz., that he had found none satisfactory as a text-book. . . . His Introduction is valuable, but his admirable Conspectus, of nearly fifty pages of fine type, gives a key to the work which will make the study of Butler a new kind of business. We have been surprised and delighted with this new aid afforded by President Malcom. In our opinion this edition will be the text-book used by all intelligent instructors who once give it an examination. Persons who have been repelled from studying Butler by the difficulties both of the argument and the style, will find these difficulties chiefly disappear by the facilities Dr. M. has afforded."—*Southern Baptist.*

Studies in the English of Bunyan. By Prof. J. B. GRIER. 12mo. Cloth. $1.25.

Phelps's Chemistry, for Beginners. Designed for Common Schools, and the Younger Pupils of Higher Schools and Academies. Revised and enlarged. By Mrs. LINCOLN PHELPS. 16mo. Half roan. 60 cents.

The design of this work is to teach Chemistry to beginners. The author has sought to present the elements of the	science in a popular and attractive form, without offending the scholar by marring its classical beauty and proportions.

Phelps's Chemistry, for Collegiate Institutions, Schools, Families, and Private Students. New edition, revised and corrected. With an Appendix containing the latest discoveries and improvements to 1866. By Mrs. LINCOLN PHELPS. 12mo. Half roan. $1.35.

Phelps's Natural Philosophy, for Schools, Families, and Private Students. By Mrs. LINCOLN PHELPS. New edition, revised and enlarged. 12mo. Half roan. $1.35.

The author has endeavored to invest the subject with freshness and interest,	to enliven the interest of the young as they climb the hill of science.

Phelps's Natural Philosophy, for Beginners. Designed for Common Schools and Families. New edition, revised and enlarged. By Mrs. LINCOLN PHELPS. 16mo. Half roan. 60 cents.

The attention of the young should be directed to natural operations, that thus	the powers of observation and comparison may be developed and strengthened.

Phelps's Lectures on Botany. Explaining the Structure, Classification, and Uses of Plants. . Illustrated upon the Linnæan and Natural Methods. New Edition. Revised and enlarged. With a Supplement, containing a Familiar Introduction to the Natural Orders, and an Artificial Key for Analysis of the same. By Mrs. LINCOLN PHELPS. The THREE HUNDRED AND SEVENTY-FIFTH THOUSAND. 12mo. Half roan. $1.60.

"The Natural System of Botany will be found here fully exhibited in all its essential features, according to the method of Prof. Lindley, and with full descriptions of Natural Orders. . . . I think your work well calculated to at-	tract beginners, and especially young ladies, to the study, by conducting them in the most agreeable way to the vestibule of the botanical temple."—HON. WM. DARLINGTON, M.D.

Frick's Physical Technics. Practical Instructions for making Experiments in Physics and the Construction of Physical Apparatus with the most limited means. By DR. T. FRICK. Translated by JOHN D. EASTER, Ph.D. Illustrated by over 800 Engravings. 8vo. Cloth. $2.50.

It is the object of this work on the one hand to furnish an introduction to physical experimentation, to describe all the particulars requisite to success, to call attention to those points which must be	considered in the purchase and use of apparatus; and on the other hand to give instructions for the construction of apparatus in the cheapest and most effective way.

Smith's Elements of the Laws; or, Outlines of
the System of Civil and Criminal Laws in Force in the United States and in the several States of the Union. Designed as a text-book and for general use. By THOMAS L. SMITH, *late Judge of the Supreme Court of the State of Indiana*. New edition, revised. 12mo. Fine cloth. $1.50.

This work is designed to enable any one to acquire a competent knowledge of his legal rights and privileges in all of the most important political and business relations of the citizens of the country, with the principles upon which they are founded, and the means of asserting and maintaining them in civil and criminal cases.

Schmitz's German Grammar. A Text-book for
the Practical Study of the German Language. By J. ADOLPH SCHMITZ, A.M., *Professor of Modern Languages and Literature in the University of Wooster, Ohio*, and HERMAN J. SCHMITZ, *Instructor of German and French, Newark, New Jersey.* 12mo. Half roan. $1.50.

"Schmitz's 'German Grammar' I have received and examined, and, to my opinion, it is far superior to any other German grammar thus far published—the best ever come to my sight.

I can fully recommend it to all teachers of the German language."—PROF. HENRY MARIN, *State Normal School, Oshkosh, Wisconsin.*

Long's Primary Grammar. An Introduction to
English Grammar. An easy Method for Beginners. By HARRIETT S. LONG, *Preceptress in English and French, St. Anna's Hall, Brookeville, Maryland.* 16mo. Boards. 25 cents.

Lieber's Civil Liberty and Self-Government. A
Treatise on Civil Liberty and Self-Government. New edition, revised and enlarged. Edited by THEODORE D. WOOLSEY. 8vo. Extra cloth. $3.15.

Lieber's Political Ethics. A Manual of Political
Ethics, designed chiefly for the use of Colleges and Students at Law. 2 vols. 8vo. Extra cloth. $5.50.

Progress of Philosophy. By Samuel Tyler,
LL.D. In the Past and in the Future. Second Edition, enlarged. 12mo. Cloth. $1.75.

Meredith's Every-Day Errors of Speech. By
L. P. MEREDITH, M.D., D.D.S., author of "The Teeth and How to Save them." 16mo. Fine cloth. 75 cents.

"If any one flatters himself that he knows how to pronounce the English language, let him buy this thin little volume of less than one hundred pages, and treat himself to a few hundred surprises. He will probably rise up from its perusal a wiser and a better man; for it treats of the common errors of speech made by educated persons, who would be glad to have their attention called to the mistakes that fall unconsciously from their tongues."—*Louisville Courier-Journal.*

Morton and Leeds's Chemistry. The Student's
Practical Chemistry. A Text-book for Colleges and Schools on Chemical Physics, including Heat, Light, and Electricity, and on Inorganic and Organic Chemistry. By HENRY MORTON, A.M., and ALBERT H. LEEDS, A.M. Illustrated with over 150 Woodcuts. 12mo. Cloth. $1.75.

In this work the student is furnished with simple and clear explanations of the subjects, and those more advanced in scientific learning with convenient memoranda of important facts, numbers, references, etc. The effort has also been made to embody all the valuable novelties in the branches discussed (many of which have not been introduced in any text-book), and thus bring this work down to the present time.

Sypher's History of Pennsylvania. From the
First Settlements on the Delaware to the Present Time. Designed for use in Schools, Academies, Colleges, Families, and Libraries. With Statistical Tables. By J. R. SYPHER. $1.25.

"We have always had good histories of the United States adapted to schools, but up to the present time no one has given us a School History of Pennsylvania. You have now supplied that want, and, in the name of the school interests of the Commonwealth, I thank you for it."—HON. J. P. WICKERSHAM, *State Supt. of Common Schools.*

"It is the best State history for common use within our knowledge, and we hope some one will compile one as good of our own State. When he does, we shall urge that it be adopted as a reader in all our common schools and seminaries, and thus brought home to every family and fireside."—*N. Y. Tribune,* editorial by HON. HORACE GREELEY.

History of New Jersey. From the Earliest Set-
tlements to the Present Time. Designed for Common Schools, Academies, Colleges, Families, and Libraries. By J. R. SYPHER and E. A. APGAR. $1.25.

"New Jersey may well be proud of her past, satisfied with her present, and hopeful as to her future; and all these are faithfully portrayed and clearly foreshadowed in this little volume. We trust it may find its way not only into the seminaries or schools, but to the families and firesides of the State."—*New York Tribune.*

"This is probably the best history of New Jersey which has yet been published. The whole story of our State is condensed into it."—*New Jersey Journal, Elizabeth.*

Cameos from English History. By the author of
"The Heir of Redclyffe." With vignette title. 12mo. Cloth. $1.25.

The endeavor has not been to chronicle facts, but to put together a series of pictures of persons and events, so as to arrest the attention and give some individuality and distinctness to the recollection by gathering together details at the most memorable moments.

Bible Gems; or, Manual of Scripture Lessons.
Specially designed for Public Schools, but equally adapted to Sunday-schools and Families. By R. E. KREMER. Illustrated. 16mo. 314 pp. Cloth. $1.00.

La Grammaire en Action. Bulwer's Lady of Lyons. With a complete Idiomatical and Grammatical Vocabulary for Translation from English into French; preceded by a Synopsis of the most useful Rules of French Grammar, and a Methodical Table of all Irregular Verbs occurring in the Text. Also, various subjects for Original Composition in French, most of them under the form (adopted at Oxford and at West Point) of Short Letters in French, to be answered in the same Language. By PROF. B. MAURICE, A.M., *formerly Assistant Professor U. S. Naval Academy.* 12mo. Extra cloth. $1.25.

*Sue's French Course. A New System of Instruc-*tion in French. By PROF. JEAN B. SUE, A.M., *late Instructor of French at the University of Pennsylvania.* Comprising:

I. First Lessons in French. An Introduction to the "Practical and Intellectual Method for Learning French." 12mo. Half roan. 90 cents.

II. A New Practical and Intellectual Method for Learning French, grounded on Nature's Teachings; adapted to the System of Noel and Chapsal. With critical remarks on Grammars used in our Schools. Half roan. $1.35.

*III. Exercises on the French Syntax; or, Prac-*tice of the New Practical and Intellectual Method for Learning French, wherein learners have to make direct application of French Rules, and rectify the deviations made from the French Syntax. 12mo. Half roan. 65 cents.

*IV. The Vicar of Wakefield. By Oliver Gold-*smith. Arranged as a Guide for the Construction of French Sentences, completing the System of the Practical and Intellectual Method for Learning French. Half roan. $1.35.

*V. A Key for the Use of Teachers and Learn-*ers, who, after completing the regular course, may further wish to prosecute their study of the French tongue. 12mo. Half roan. 65 cents.

"Mr. Sue's New Method of imparting the French language is undoubtedly the best that has yet appeared."—MAD. CAROLINE CORSON.

"It is, in my opinion, the best practical treatise ever published, to teach the French language to English or Anglo-American pupils."—PROF. CHAS. O. DE JUVILLE, *West Chester, Pa.*

"I am acquainted with no other published course that appears to me so well calculated by its method, by its clearness and precision of its rules, and by the appropriateness of its exercises, to ground the pupil solidly in an accurate and familiar knowledge of the language."— GEO. ALLEN, *Prof. of Greek and Latin in the University of Pennsylvania.*

Fenelon, M. De. Les Aventures de Télémaque. D'après l'Edition de M. Chas. Le Brun. 12mo. Half roan. $1.25.

SANFORD'S
SERIES OF ARITHMETICS.
SANFORD'S ANALYTICAL SERIES.
COMPRISED IN FOUR BOOKS.

The Science of Numbers reduced to its last analysis. Mental and Written Arithmetic successfully combined in each Book of the Series.

By SHELTON P. SANFORD, A.M.,
Professor of Mathematics in Mercer University, Georgia.

FIRST BOOK.
Sanford's First Lessons in Analytical Arithmetic. Comprising Mental and Written Exercises. Handsomely and appropriately Illustrated. 16mo. Half roan. 27 cents.

SECOND BOOK.
Sanford's Intermediate Analytical Arithmetic. Comprising Mental and Written Exercises. 16mo. 232 pp. Half roan. 45 cents.

THIRD BOOK.
Sanford's Common School Analytical Arithmetic. 12mo. 355 pp. Half roan. 80 cents.

FOURTH BOOK.
Sanford's Higher Analytical Arithmetic; or, THE METHOD of making ARITHMETICAL CALCULATIONS on PRINCIPLES of UNIVERSAL APPLICATION, without the AID of FORMAL RULES. 12mo. 419 pp. Half roan. Cloth sides. $1.25.

From Prof. HUGH S. THOMPSON, *Principal Columbia Male Academy, Columbia, S. C.*

"Sanford's Arithmetics are superior to any that I have seen in the fulness of the examples, the clearness and simplicity of the analyses, and the accuracy of the rules and definitions. This opinion is based upon a *full* and *thorough test* in the school-room. To those teachers who may examine these Arithmetics with reference to introduction, I would especially commend the treatment of Percentage and Profit and Loss. No text-books that I have ever used are so satisfactory to teachers and pupils."

From Capt. S. Y. CALDWELL, *Supt. of Nashville (Tenn.) Public Schools.*

"Your Intermediate and Advanced Analytical Arithmetics are among the best I have examined.

"It is contrary to my practice to write testimonials or recommendations, but the high merit of your book certainly justifies it in this instance."

From Prof. B. MALLON, *Superintendent of Atlanta (Ga.) Public Schools.*

"I think they [Sanford's Arithmetics] are the best books on the subject ever published; and I trust it will not be long before they will be introduced into every school in our State. In my judgment they are the very perfection of school-books on Arithmetic."

BOOKS FOR TEACHERS.

I.
WICKERSHAM'S METHODS OF INSTRUCTION;

OR,

That Part of the Philosophy of Education which Treats of the Nature of the Several Branches of Knowledge and the Method of Teaching Them.

By J. P. WICKERSHAM, A.M.,

State Superintendent of Public Instruction of Pennsylvania.

12mo. Cloth. $1.75.

II.
WICKERSHAM'S SCHOOL ECONOMY.

A Treatise on the Preparation, Organization, Employments, Government, and Authorities of Schools.

By J. P. WICKERSHAM, A.M.,

State Superintendent of Public Instruction of Pennsylvania.

12mo. Cloth. $1.50.

SPECIAL CHARACTERISTICS.

I. These works are a philosophical exposition of that part of education of which they treat. Every division will be found in its proper place, and good reasons are always given for its statements.

II. They are practical. Every teacher can make an application of their principles. They are especially valuable as guides to young teachers.

III. Their style is clear and pointed. No rambling discussions, loose narratives, or nonsensical stories will be found within their pages. They claim rank with the more sober and solid treatises which form the standard works on law and medicine.

IV. They are exhaustive. Matter scattered through dozens of volumes on teaching is brought together and condensed in these, and nothing of importance appertaining to the subject is omitted.

V. They are now used as text-books with marked success in a number of State Normal Schools, Private Normal Schools, Teachers' Institutes and Associations.

VI. They contain matter which every parent and every school officer as well as every teacher should be acquainted with.

PUBLICATIONS OF J. B. LIPPINCOTT & CO.

A NEW HISTORY OF THE UNITED STATES.

LEEDS'S
UNITED STATES HISTORY.

A History of the United States of America. Including some Important Facts mostly omitted in the Smaller Histories. Designed for General Reading and for Academies.

BY JOSIAH W. LEEDS.

Revised Edition   With Maps.   12mo.   Extra Cloth. $1.75.

"An honest and truthful book, and worthy of welcome acceptation by all who can appreciate the warp and woof of American history in their true texture and strength, without the gloss of a partisanship or patriotism that can see only one side of a question or one aspect of a fact. . . . No other volume of the same size could contain more extensive or varied information, or classify it in better proportioned departments. As it stands, it is the only complete history of our country from the discovery of the Northmen to the election of Mr. Hayes, embracing all the leading events between these two widely-separated dates."—*Literary World* (Boston).

"We are prepared to speak of it in high terms of commendation. The work is not sectional. . . . An air of calmness and candor pervades the book."—*Nashville Christian Advocate.*

"We would heartily commend it as being, in many respects, the best United States History at present extant for use in schools, and for a place in the library or the family."—*Friends' Review.*

DERRY'S
HISTORY OF THE UNITED STATES.
FOR SCHOOLS AND ACADEMIES.

By JOSEPH T. DERRY,

Professor of Ancient Languages in the Academy of Richmond Co., Augusta, Ga.

With Numerous Illustrations.   12mo.   Extra Cloth. $1.35.

This work is comprehensive, embracing the most important facts in the history of our country.

Its style is perspicuous. The author's aim has been to place important facts within easy comprehension of any child old enough to take up the study of history. Derry's History is designed as a text-book for schools, and therefore avoids metaphysical discussions of the theory of our government. That portion which treats of the formation of the Constitution and the establishment of the Republic is easily understood.

From Rev. W. W. HENDRIX, *President Bethel College, Tenn.*

"Having examined this (Derry's) History, and used it in the class-room, I am free to recommend it as very well suited for the purpose it is intended to fill. It is simply full of *history*, without the rubbish of individual opinions and partisan prejudices." March 30, 1876.

From Rev. ATTICUS G. HAYGOOD, D.D., *President Emory College, Oxford, Ga.*

"Prof. Derry has given us a History for schools and students in the form of alternate question and answer. We believe that he has 'hit the mark.' Prof. D. has done his work thoroughly, conscientiously. It is the book, as it seems to us, for Southern schools, as it is just to both sections."

PUBLICATIONS OF J. B. LIPPINCOTT & CO.

CUTTER'S SERIES
ON
Analytical Anatomy, Physiology, and Hygiene,

HUMAN AND COMPARATIVE,

FOR SCHOOLS AND FAMILIES.

By CALVIN CUTTER, A.M., M.D.

During the past ten years more than two hundred thousand (200,000) have been sold for schools. This is the only series of works upon the subject that is graded for all classes of pupils, from the primary school to the college, the only one that embraces Anatomy, Physiology, and Hygiene for schools, and the only one arranged so as to be used advantageously with illustrating Anatomical Charts.

NEW SERIES.

First Book on Analytic Anatomy, Physiology, and Hygiene, Human and Comparative. 196 pp. With 164 Illustrations. 12mo. 80 cents.

Second Book on Analytic Anatomy, Physiology, and Hygiene, Human and Comparative. With QUESTIONS, DIAGRAMS, AND ILLUSTRATIONS for Analytic Study and Unific Topical Review. 309 pp. With 186 Illustrations. 12mo. $1.35.

New Analytic Anatomy, Physiology, and Hygiene, Human and Comparative. With QUESTIONS, DIAGRAMS, AND ILLUSTRATIONS for Analytic Study and Synthetic Review. 388 pp. With 230 Illustrations. 12mo. $1.50.

OLD SERIES.

First Book on Anatomy, Physiology, and Hygiene. Illustrated. 12mo. 70 cents.

Anatomy, Physiology, and Hygiene. Illustrated. 12mo. $1.50.

Human and Comparative Anatomy, Physiology, and Hygiene. By Mrs. E. P. CUTTER. Illustrated. 12mo. 45 cents.

NEW OUTLINE ZOOLOGICAL CHARTS,

Or **Human and Comparative Anatomical Plates.** Mounted. (Including Rollers, etc.) Nine in number. Three feet long by two feet wide. $15.00. Half set. Five in number. $9.00.

www.ingramcontent.com/pod-product-compliance
Lightning Source LLC
Chambersburg PA
CBHW020823230426
43666CB00007B/1085